Medical Microbiology and Infection
at a Glance

Medical Microbiology and Infection at a Glance

Stephen H. Gillespie

MD, DSc, FRCP (Edin), FRCPath
Professor of Medical Microbiology and Regional Microbiologist
Royal Free and University College Medical School
(Royal Free Campus)
University College London
London

Kathleen B. Bamford

MD, FRCPath
Consultant Medical Microbiologist and Visiting Professor
Imperial College London
Hammersmith Hospital
London

Third edition

© 2007 Stephen H. Gillespie, Kathleen B. Bamford
Published by Blackwell Publishing
Blackwell Publishing, Inc., 350 Main Street, Malden, Massachusetts 02148-5020, USA
Blackwell Publishing Ltd, 9600 Garsington Road, Oxford OX4 2DQ, UK
Blackwell Publishing Asia Pty Ltd, 550 Swanston Street, Carlton, Victoria 3053, Australia

First published 2000
Second edition 2003
Third edition 2007

1   2007

Library of Congress Cataloging-in-Publication Data

Gillespie, S. H.
    Medical microbiology and infection at a glance / by Stephen H. Gillespie, Kathleen
B. Bamford. — 3rd ed.
        p. ; cm. — (At a glance)
    Includes index.
    ISBN-13: 978-1-4051-5255-6
    ISBN-10: 1-4051-5255-9
    1. Medical microbiology.   I. Bamford, Kathleen B.   II. Title.   III. Series: At a glance series
(Oxford, England)
    [DNLM: 1. Microbiology.   2. Communicable Diseases.   3. Infection.   QW 4 G478m 2007]

    QR46.G47 2007
    616.9′041—dc22

                                                           2007017557

    ISBN: 978-1-4051-5255-6

A catalogue record for this title is available from the British Library

Set in 9.5/12pt Times by Graphicraft Limited, Hong Kong
Printed and bound in Singapore by Fabulous Printers Pte Ltd

Commissioning Editor: Vicki Donald
Editorial Assistant: Robin Harries
Development Editor: Beckie Brand
Production Controller: Debbie Wyer

For further information on Blackwell Publishing, visit our website:
http://www.blackwellpublishing.com

Contents

Preface to the third edition

Infectious diseases and microbiology is a discipline that moves rapidly. Not only does medical science advance, but the organisms respond to changes in the environment and our medical interventions. New organisms continue to be discovered and new treatments, especially in the area of anti-viral chemotherapy, have been developed. The way in which we diagnose infections is now also changing rapidly with an increasing number of molecular diagnostic assays becoming available. The authors have endeavoured to incorporate this changing biological and medical landscape into this edition of *Medical Microbiology and Infection at a Glance*.

In the last few years there has been an increased recognition of the importance of infection acquired in hospital. This has occurred, in part, as a result of the emergence of multiple drug resistance organisms such as glycopeptide resistant *Staphylococcus aureus* and the increase in the incidence and severity of *Clostridium difficile* infections. We have responded to this by increasing our emphasis on this subject and adding a chapter on vaccination. We have also added a chapter on streptococcal infection.

At the end of the book we have added a self-assessment section containing some cases that give examples of how infectious diseases present are diagnosed and treated.

It is clear that infectious diseases will continue to be an important subject for medical students, doctors and other professionals in all areas of healthcare. The authors hope that his new, updated edition will help its readers to respond to this threat to our patients.

Stephen H. Gillespie & Kathleen B. Bamford
London 2007

Preface to the first edition

This book is written for medical students and doctors who are seeking a brief summary of microbiology and infectious diseases. It should prove useful to those embarking on a course of study and assist those preparing for professional examinations.

Chapters are divided into concepts, the main human pathogens and the infectious syndromes. This broadly reflects the pattern of teaching in many medical schools.

Microbiology is a rapidly growing and changing subject: new organisms are constantly being identified and our understanding of the pathogenic potential of recognized pathogens is being expanded. In addition the taxonomists keep changing the names of familiar friends to add to the confusion. Despite this, there are clear fundamental facts and principles that form a firm foundation of knowledge on which to build throughout a professional career. It is these that this book strives to encapsulate.

Each chapter contains a diagram which illustrates core knowledge. The associated text offers further insights and details where necessary.

Irrespective of a doctor's specialty, diligent study of microbiology provides the basis for sound professional judgement, giving the clinician confidence and benefiting patients for years to come.

The authors gratefully acknowledge the editorial work of Dr Janet Gillespie who has reminded the authors of practice in a community setting. They are also grateful to Dr Deenan Pillay for his critical reading of the virology sections.

Stephen Gillespie & Kathleen Bamford
London, 2000

1 Structure and classification of bacteria

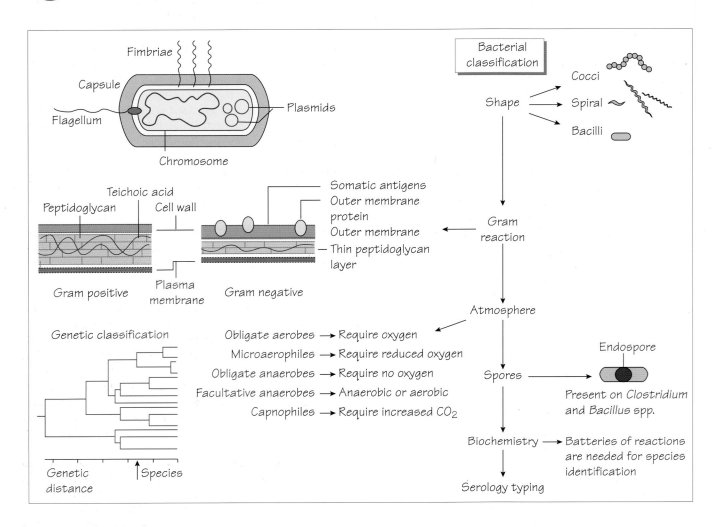

Bacterial structural components

The rigid bacterial cell wall maintains its shape and protects the cell from differences in osmotic tension between the cell and the environment. Gram-positive cell walls have a thick peptidoglycan layer and a cell membrane, whereas Gram-negative cell walls have three layers: an inner and outer membrane and a thinner peptidoglycan layer. The mycobacterial cell wall has a high proportion of lipid, including immunoreactive antigens. Bacterial shape is used in classification: cocci are spherical; bacilli are long and thin, with coccobacilli in between; and there are also curved and spiral bacilli with different wavelengths. Important cell structures include the following.

- **Capsule**: a loose polysaccharide structure protecting the cell from phagocytosis and desiccation.
- **Lipopolysaccharide**: protects Gram-negative bacteria from complement-mediated lysis. A potent stimulator of cytokine release.
- **Fimbriae** or **pili**: specialized thin projections that aid adhesion to host cells and colonization. Uropathogenic *Escherichia*

coli have specialized fimbriae (P fimbriae) that bind to mannose receptors on ureteric epithelial cells. Fimbrial antigens are often immunogenic but vary between strains so that repeated infections may occur (e.g. *Neisseria gonorrhoeae*).

- **Flagella**: bacterial organs of locomotion, enabling organisms to find sources of nutrition and penetrate host mucus. Flagella can be single or multiple, at one end of the cell (polar) or at many points (peritrichous). In some species (e.g. *Treponema*), the flagella are firmly fixed within the bacterial cell wall.
- **Slime**: polysaccharide material secreted by some bacteria growing in biofilms that protects the organism against immune attack and eradication by antibiotics.
- **Spores**: a metabolically inert form triggered by adverse environmental conditions; adapted for long-term survival, allowing regrowth under suitable conditions.

Bacteria are prokaryotes, that is they have a single chromosome and lack a nucleus. To pack the chromosome inside the cell the DNA is coiled and supercoiled; a process mediated by the DNA gyrase enzyme system (see Chapter 6). Bacterial ribosomes differ from eukaryotic ones, making them a target for antibacterial therapy. Bacteria also contain accessory DNA in the form of plasmids. For the role of plasmids in antimicrobial resistance see Chapter 7. They may also code for pathogenicity factors.

Classification of bacteria

The purpose of classification of microorganisms is to define the pathogenic potential. For example, a *Staphylococcus aureus* isolated from blood is more likely to be acting as a pathogen than *Staphylococcus epidermidis* from the same site. Some bacteria have the capacity to spread widely in the community and cause serious disease, for example *Corynebacterium diphtheriae* and *Vibrio cholerae*. Bacteria are identified using a series of physical immunological or molecular characteristics.

• **Gram reaction**: Gram-positive and Gram-negative bacteria respond to different antibiotics. Other bacteria (e.g. mycobacteria) may require special staining techniques.

• **Cell shape** (cocci, bacilli or spirals).

• **Endospore**: presence, shape and position in the bacterial cell (terminal, subterminal or central).

• **Atmospheric preference**: aerobic organisms require oxygen; anaerobic ones require an atmosphere with very little or no oxygen. Organisms that grow in either atmosphere are known as facultative anaerobes. Microaerophiles prefer a reduced oxygen tension; capnophiles prefer increased carbon dioxide.

• **Fastidiousness**: requirement for special media or intracellular growth.

• **Key enzymes**: for example, lack of lactose fermentation helps identify salmonellae, urease helps identify *Helicobacter*.

• **Serological reactions**: interaction of antibodies with surface structures (e.g. subtypes of salmonellae, *Haemophilus*, meningococcus and many others).

• **DNA sequences**: 16S ribosomal DNA sequences are now a key element in classification.

The classification systems used are very effective, but it is important to remember that these are generalizations and that there can be considerable variation in clinical behaviour of different strains of bacteria within a species as well as similarities across species. For example, some strains of *E. coli* may cause similar diseases to *Shigella sonnei*, and toxin-producing *C. diphtheriae* causes different disease from non-toxin producers.

Medically important groups of bacteria

Gram-positive cocci are divided into two main groups: the staphylococci (catalase-positive), for example the major pathogen *Staphylococcus aureus*; and the streptococci (catalase-negative), for example the major pathogens *Streptococcus pyogenes*, an agent of sore throat and rheumatic fever, and *Streptococcus agalactiae*, a cause of neonatal meningitis and pneumonia (see Chapters 14 and 15).

Gram-negative cocci include the pathogenic *Neisseria meningitidis*, an important cause of meningitis and septicaemia, and *N. gonorrhoeae*, the agent of urethritis (gonorrhoea).

Gram-negative coccobacilli include the respiratory pathogens *Haemophilus* and *Bordetella* (see Chapter 20) and zoonotic agents, such as *Brucella* and *Pasteurella* (see Chapter 21).

Gram-positive bacilli are divided into sporing and non-sporing. The sporing are subdivided between those that are aerobic (*Bacillus*: see Chapter 16) and those that are anaerobic (*Clostridium*: see Chapter 18). Pathogens include *Bacillus anthracis* which causes anthrax, and clostridia which cause gas gangrene, tetanus, pseudomembranous colitis and botulism. Non-sporing pathogens include *Listeria* and corynebacteria (see Chapter 16).

Gram-negative bacilli, including the facultative family Enterobacteriaceae, form part of the normal flora of humans and animals and can be found in the environment. They include many pathogenic genera: *Salmonella*, *Shigella*, *Escherichia*, *Proteus* and *Yersinia* (see Chapter 23). *Pseudomonas*, an environmental saprophyte naturally resistant to antibiotics, has become an important hospital pathogen (see Chapter 25). *Legionella* is another environmental species that lives in water but causes human infection if conditions allow (see Chapter 25).

Spiral bacteria include the small gastrointestinal pathogen *Helicobacter* that colonizes the stomach, leading to gastric and duodenal ulcer and gastric cancer, and *Campylobacter* spp. that cause acute diarrhoea (see Chapter 27). The *Borrelia* give rise to relapsing fever (*B. duttoni* and *B. recurrentis*) and to a chronic disease of the skin joints and central nervous system, Lyme disease (*B. burgdorferi*). The *Leptospira* are zoonotic agents causing an acute meningitis syndrome that may be accompanied by renal failure and hepatitis. The *Treponema* include the causative agent of syphilis (*T. pallidum*).

Rickettsia, Chlamydia and *Mycoplasma*

Of these, only *Mycoplasma* can be isolated on artificial media; the others require isolation in cell culture, or diagnosis by molecular or serological techniques.

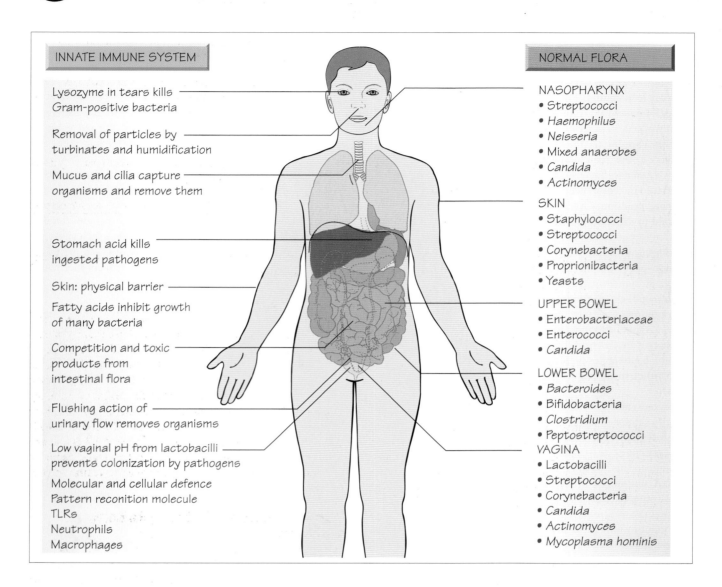

INNATE IMMUNE SYSTEM

Lysozyme in tears kills Gram-positive bacteria

Removal of particles by turbinates and humidification

Mucus and cilia capture organisms and remove them

Stomach acid kills ingested pathogens

Skin: physical barrier

Fatty acids inhibit growth of many bacteria

Competition and toxic products from intestinal flora

Flushing action of urinary flow removes organisms

Low vaginal pH from lactobacilli prevents colonization by pathogens

Molecular and cellular defence
Pattern reconition molecule
TLRs
Neutrophils
Macrophages

NORMAL FLORA

NASOPHARYNX
• Streptococci
• Haemophilus
• Neisseria
• Mixed anaerobes
• Candida
• Actinomyces

SKIN
• Staphylococci
• Streptococci
• Corynebacteria
• Proprionibacteria
• Yeasts

UPPER BOWEL
• Enterobacteriaceae
• Enterococci
• Candida

LOWER BOWEL
• Bacteroides
• Bifidobacteria
• Clostridium
• Peptostreptococci

VAGINA
• Lactobacilli
• Streptococci
• Corynebacteria
• Candida
• Actinomyces
• Mycoplasma hominis

The most important defence against infection is the innate immune system, which consists of the normal flora, physical barriers like the skin, antibacterial proteins and phagocytic cells. Many responses to 'harm' are recognized by pattern recognition molecules such as the Toll-like receptors (TLRs) which, when engaged by bacterial components, trigger cascades that activate phagocytes and the immune response. For example, TLR-4 recognizes lipopolysaccharide and TLR-9 recognizes cpgDNA. The main components of the system are listed in Table 2.1. Variation in the expression/composition of each component affects an individual's resistance to infection.

Normal flora

There are many more prokaryotic (bacterial) than human cells in the body. This normal flora protects by competing with pathogens for colonization sites and producing antibiotic substances (bacteriocins) that suppress competing organisms. Anaerobic bacteria produce toxic metabolic products and free fatty acids that inhibit other organisms. In the female genital tract lactobacilli produce lactic acid that lowers the pH, preventing colonization by pathogens.

Antibiotics suppress normal flora, allowing colonization and infection by naturally resistant organisms, such as *Candida albicans*. The infective dose of *Salmonella typhi* is lowered by concomitant antibiotic use. Antibiotics may upset the balance between organisms of the normal flora, allowing one to proliferate disproportionately, for example *Clostridium difficile* infection, resulting in a severe diarrhoeal disease (Chapter 18).

Physical and chemical barriers

The skin provides a physical barrier to invasion, secreting sebum and fatty acids that inhibit bacterial growth. Many organisms

Table 2.1 The innate immune system: the location of barriers to infection and the mechanisms and consequences of deficiency.

Component	Compromise	Consequence
Normal flora		
Pharynx	Antibiotics	Oral thrush
Intestine	Antibiotics	Pseudomembranous colitis; colonization with antibiotic-resistant organisms
Vagina	Antibiotics	Vaginal thrush
Skin	Burns, vectors	Cutaneous bacterial infection, infection with pathogenic viruses, bacteria, protozoa and metazoa
Turbinates and mucociliary clearance	Kartagener's syndrome, cystic fibrosis, bronchiectasis	Chronic bacterial infection
Lysozyme in tears	Sjögren's syndrome	Ocular infection
Urinary flushing	Obstruction	Recurrent urinary infection
Phagocytes, neutrophils, macrophages	Congenital, iatrogenic, infective	Chronic pyogenic infection, increased susceptibility to bacterial infection
Complement	Congenital deficiency	Increased susceptibility to bacterial infection, especially *Neisseria* and *Streptococcus pneumoniae*

have evolved mechanisms to penetrate the skin, either via the bite of a vector, e.g. *Aedes aegypti* bite transmitting dengue, or by invasion through intact skin, e.g. *Leptospira* and *Treponema*. Some organisms colonize mucosal surfaces and use this route to gain access to the body.

Breaking skin integrity by intravenous cannulation, or medical or non-medical injection, can transmit blood-borne viruses such as hepatitis B or HIV. Diseases of the skin, such as eczema or burns, permit colonization and invasion by pathogens (e.g. *Streptococcus pyogenes*).

Mucociliary clearance mechanism In the respiratory tract air is humidified and warmed by passage over the turbinate bones and through the nasal sinuses. Particles settle on the sticky mucus of the respiratory epithelium and the debris is transported by the cilial 'conveyor belt' to the oropharynx where it is swallowed. This allows only particles less than 5 µm diameter to reach the alveoli: the respiratory tract is effectively sterile below the carina.

Secreted antibacterial compounds Mucus contains polysaccharides of similar antigenic structure to the underlying mucosal surface; organisms bind to the mucus and are removed. The body secretes antibacterial compounds—for example, lysozyme in tears degrades Gram-positive bacterial peptidoglycan; lactoferrin in breast milk binds iron, inhibiting bacterial growth; lactoperoxidase, a leucocyte enzyme, produces superoxide radicals that are toxic to microorganisms.

Gastric acid protects from intestinal pathogens—acid suppression increases the risk of intestinal infection.

Urinary flushing The urinary tract is protected by the flushing action of urinary flow and, except near the urethral meatus, is sterile. Obstruction by stones or tumours, benign prostatic hypertrophy or scarring of the urethra or bladder may cause reduction of the urinary flow and stasis, with subsequent bacterial urinary infection.

Phagocytes

Neutrophils and macrophages ingest particles, including bacteria, viruses and fungi. Opsonins (e.g. complement and antibody) may enhance phagocytic ability. For example, *Streptococcus pneumoniae* are not phagocytosed unless their capsule is coated with an anticapsular antibody. The action of macrophages in the reticuloendothelial system is essential for resistance to many bacterial and protozoan pathogens, for example *S. pneumoniae* and malaria. Congenital deficiency in neutrophil function leads to chronic pyogenic infections, recurrent chest infections and bronchiectasis. Following splenectomy, patients have defective macrophage function and diminished ability to remove capsulate organisms from the blood.

Complement and other plasma proteins

Complement is a system of plasma proteins that collaborate to resist bacterial infection. The complement cascade is activated by antigen–antibody binding (the classical pathway) or by direct interaction with bacterial cell wall components (the alternative pathway). The products of both processes attract phagocytes to the site of infection (chemotaxis), activate phagocytes, cause vasodilatation and stimulate phagocytosis of bacteria (opsonization). The final three components of the cascade form a 'membrane attack complex' that can lyse Gram-negative bacteria. Complement deficiencies render patients susceptible to acute pyogenic infections, especially with *Neisseria meningitidis*, *N. gonorrhoeae* and *Streptococcus pneumoniae*.

Transferrin is a transport vehicle for iron, limiting the amount of iron available to invading microorganisms. Other acute-phase proteins are directly antibacterial, for example mannose-binding protein or C-reactive protein (CRP), which bind to bacteria and activate complement.

3 Pathogenicity and pathogenesis of infectious disease

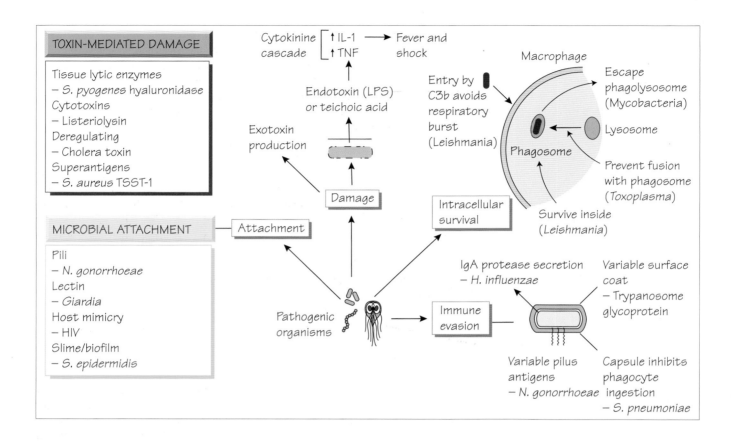

TOXIN-MEDIATED DAMAGE

Tissue lytic enzymes
- S. pyogenes hyaluronidase
Cytotoxins
- Listeriolysin
Deregulating
- Cholera toxin
Superantigens
- S. aureus TSST-1

MICROBIAL ATTACHMENT

Pili
- N. gonorrhoeae
Lectin
- Giardia
Host mimicry
- HIV
Slime/biofilm
- S. epidermidis

Cytokinine cascade ↑ IL-1 → Fever and shock
↑ TNF

Endotoxin (LPS) or teichoic acid

Exotoxin production

Damage

Attachment

Pathogenic organisms

Immune evasion

Intracellular survival

Macrophage

Entry by C3b avoids respiratory burst (Leishmania)

Phagosome

Escape phagolysosome (Mycobacteria)

Lysosome

Prevent fusion with phagosome (Toxoplasma)

Survive inside (Leishmania)

IgA protease secretion
- H. influenzae

Variable surface coat
- Trypanosome glycoprotein

Variable pilus antigens
- N. gonorrhoeae

Capsule inhibits phagocyte ingestion
- S. pneumoniae

Definitions

The normal host is colonized by bacteria and protozoa that do not cause disease. An infection occurs when microorganisms cause ill-health. They may do this by invading host tissues or by exerting effects from mucosal surfaces. An organism capable of causing an infection is a pathogen; one which forms part of the normal flora is a commensal. Pathogenicity is the capacity to cause disease; whereas virulence is the ability to cause serious disease. For example, the main pathogenicity determinant of *Streptococcus pneumoniae* is the capsule, without which it cannot cause disease. Some capsular types cause more serious disease: they alter the virulence (Chapter 15). The term parasite is often used to describe protozoan and metazoan organisms, but this is confusing as these organisms are either pathogens or commensals.

Types of pathogen

Obligate pathogens are almost always associated with disease (e.g. *Treponema pallidum*, HIV). Conditional pathogens may cause disease if certain conditions are met. For example, *Bacteroides fragilis* is a normal commensal of the gut but if it reaches

the peritoneal cavity, particularly together with coliform-type bacteria, it will cause abscesses; *Staphylococcus aureus* is a commensal of the anterior nares that may cause disease if inoculated into a wound. Other organisms are opportunistic pathogens. These usually only affect immunocompromised hosts. For example, *Pneumocystis jiroveci* usually only causes lung infection in a host that has severely compromised T-cell immunity.

Mechanisms of disease

For an organism to cause disease it must satisfy a number of criteria.

Access to a vulnerable host — transmission

Different organisms are transmitted by different means. In some cases epidemic strains may be more efficiently transmitted or may be able to survive the rigors of interhost transmission more effectively, thus spreading more rapidly. Many pathogens are adapted to particular routes of transmission (Chapter 8). It is also worth noting that respiratory pathogens induce coughing and therefore facilitate spread by respiratory droplet. Likewise organisms that are transmitted by the faecal–oral route induce vomiting and diarrhoea, and thus contaminate the environment by being present in high numbers in gastrointestinal secretions.

Attachment to the host

Microorganisms must attach themselves to host tissues to colonize the body; different microorganisms have different strategies and mechanisms for attaching to host tissues. The distribution of receptors that a particular organism can interact with will define the organs that are involved. *Neisseria gonorrhoeae* adheres to the genital mucosa using fimbriae. Influenza virus attaches to host cells by its haemagglutinin antigen. This accounts for both species-specific pathogenesis (e.g. the ability of certain strains to cause disease in different species such as avian or porcine strains) and intraspecies variation in affinity and susceptibility.

Some bacteria have mechanisms that help them get close to the mammalian epithelium. For example, *Vibrio cholerae* excretes a mucinase to help it reach the enterocyte. By varying the sialation of surface structures, *Helicobacter pylori* interacts differently with sialated proteins in gastric mucus or cell surfaces. *Giardia lamblia* attaches to the jejunal mucosa by a specialized sucking disc. Red cells infected with *Plasmodium falciparum* express a parasite-encoded protein that mediates adherence to host brain capillaries and is responsible for cerebral malaria.

Some bacteria form a polysaccharide biofilm that aids colonization of indwelling prosthetic devices, such as catheters. Some strains of staphylococci have genes that mediate attachment to plastics and to biological molecules that coat intravascular devices.

Different strains of *S. aureus* can adhere or bind to a variety of host molecules that may be exposed following tissue damage, such as fibronectin, vibronectin and collagen; different strains of *Escherichia coli* express fimbriae or pili associated with adhesions involving mannose (mannose-binding proteins) or P blood group antigens, and are associated with gastrointestinal and urinary tract infection respectively; HIV binds to CD4 antigen and a number of others. Many of the genes involved in the attachment ability of organisms are present or expressed variably across strains.

Invasion

Microorganisms have a variety of strategies that allow them to cross mucosal barriers or different types of cell membrane. Once they have crossed this barrier they must then survive and multiply when they invade the host. Some bacteria such as *Helicobacter* and *Neisseria* produce IgA proteases. These enzymes break down IgA and thus are able to overcome one of the cardinal mucosal defence systems.

Motility

The ability to move in order to locate new sources of food or in response to chemotactic signals potentially enhances pathogenicity. *Vibrio cholerae* is motile by virtue of its flagellum—non-motile mutants are less virulent.

Immune evasion

To survive in the human host, pathogens must overcome the host immune defence. Respiratory bacteria secrete an IgA protease that degrades host immunoglobulin. *Staphylococcus aureus* expresses protein A which binds host immunoglobulin, preventing opsonization and complement activation.

Avoiding destruction by host phagocytes is an important evasive technique. *Streptococcus pneumoniae* has a polysaccharide capsule that inhibits uptake by polymorphonuclear neutrophils (PMNs). Some organisms are specially adapted to survive inside host macrophages, for example *Toxoplasma gondii*, *Leishmania donovani* and *Mycobacterium tuberculosis*. The lipopolysaccharide (LPS) of Gram-negative organisms makes them resistant to the effect of complement. *Trypanosoma* alter surface antigens to evade antibodies.

Damaging the host
Toxins
Endotoxins

Endotoxins stimulate macrophages to produce interleukin-1 (IL-1) and tumour necrosis factor (TNF), causing fever and shock.

Exotoxins

Some organisms secrete exotoxins that cause local or distant damage, these are usually proteins. Many have a subunit structure. Often one type of subunit facilitates attachment or entry to host cells while another mediates the physiological effects. Cholera toxin is a classic example where the B subunit binds to the epithelial cell and the A subunit activates adenyl cyclase resulting in sodium and chloride efflux from the cell, thus causing diarrhoea.

Other exotoxins act as superantigens, causing non-specific activation of T cells with compatible variable region structure causing inflammatory cytokine production, which in turn results in widespread physiological effects with fever, shock, gastrointestinal disturbance and rash.

Some exotoxins interfere with host cell protein synthesis (e.g. diphtheria toxin and *P. aeruginosa* exotoxin A), others inferfere with neurological or neuromuscular signalling (e.g. tetanus and botulinum toxin).

In many cases antibody to the toxin ameliorates the physiological effects of the disease and is therefore protective (see Chapter 11).

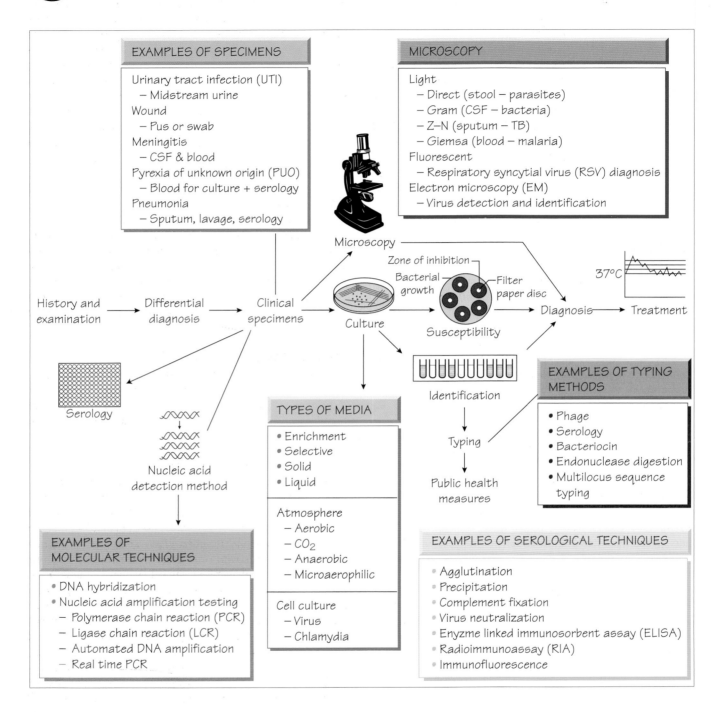

Specimens

Any tissue or body fluid can be subjected to microbiological investigation. Culture is used to increase the number of bacteria present to permit identification and susceptibility testing. This may require the use of enrichment media. In specimens with a normal flora, it is necessary to inhibit the non-pathogens and encourage the growth of pathogens: selective media are required. Appropriate specimen collection is critical to achieving a useful result; for example, poor aseptic technique leading to contamination of blood cultures may result in inappropriate therapy.

Many bacteria do not survive well outside the body: obligate anaerobes may be killed by atmospheric oxygen. Some organisms are very susceptible to drying (e.g. *Neisseria gonorrhoeae*). To protect organisms during transportation the specimen may be plated onto a suitable medium immediately or inoculated into a transport medium.

Laboratory examination

Specimens may be examined directly, for example to detect the presence of adult worms in faeces or blood in sputum. Microscopic examination is rapid and demands little expensive equipment, but requires considerable technical expertise and is insensitive: a large number of organisms must be present to achieve a positive diagnosis. It also lacks specificity because commensal organisms may be mistaken for pathogens.

Special stains can be used to identify organisms (e.g. auramine or Ziehl–Neelsen's (Z–N) method for mycobacteria). Silver methenamine stains the chitin in the cell wall of fungi and *Pneumocystis jiroveci*. Giemsa is useful for staining malaria and other parasites, such as *Leishmania*.

Immunofluorescence uses antibodies specific to a pathogen that are labelled with a fluorescent marker. The presence of the pathogen is confirmed when examined under ultraviolet light, as bound antibody glows as a bright apple-green fluorescence.

Culture

Even when causing severe symptoms, the infecting organism may be present in numbers that are too low to be detected by direct microscopy. Culture amplifies the number of organisms.

Culture takes two forms: growth in liquid medium amplifies the number of organisms present; growth on solid media produces individual colonies that can be separated for identification, susceptibility testing and typing. Most human pathogens are fastidious, requiring media supplemented with peptides, sugars and nucleic acid precursors (present in blood or serum). An appropriate atmosphere must also be provided: fastidious anaerobes require an oxygen-free atmosphere whereas strict aerobes such as *Bordetella pertussis* require the opposite. Most human pathogens are incubated at 37°C, although some fungal cultures are incubated at 30°C.

Identification

Different organisms cause different disease syndromes. Identifying the organism can often predict the clinical course: *Vibrio cholerae* causes a different spectrum of symptoms from *Shigella sonnei*. Identification of certain organisms may lead to public health action, for example the isolation of *Neisseria meningitidis* from cerebrospinal fluid (CSF).

Identification is based on colonial morphology on agar, the Gram stain, the presence of spores and simple biochemical tests, such as catalase or coagulase. Precise species identification usually depends on the results of a series of biochemical tests (e.g. urease activity), or the detection of bacterial products (e.g. indole). Organisms that are difficult or impossible to grow can be identified by DNA amplification techniques and sequencing (e.g. *Trophyrema whippelii*).

Susceptibility testing

Organisms are defined as susceptible if a normal dose of an antimicrobial is likely to result in cure, moderately resistant if cure is likely with a larger dose, and resistant if antibiotic therapy is likely to fail. A variety of methods can be used to determine susceptibility. The British Society of Antimicrobial Chemotherapy (BSAC) and the Clinical Laboratory Standards Institute (CLSI) methods use standardized conditions to differentiate between sensitive and resistant bacteria based on the diameter of the zone of inhibition of growth by an antimicrobial. The minimum inhibitory concentration can be measured using E-tests, by broth dilution or agar incorporation. Paper discs impregnated with antibiotic are placed on agar inoculated with the test organism. The antibiotic diffuses into the surrounding agar and inhibits bacterial growth. The extent of this inhibition reflects the susceptibility of the organism. Clinical response depends on host factors, and thus *in vitro* tests only provide an approximate guide to therapy.

Serology

Infection can be diagnosed by detecting the immune response to a pathogen. Different methods are used, including agglutination, complement fixation, virus neutralization and enzyme immunoassay (EIA). A diagnosis is made by detecting rising or falling antibody levels in specimens obtained more than a week apart, or the presence of specific IgM. Alternatively, specific antigen can be detected; for example, agglutination techniques can be used to detect bacterial capsular antigens in CSF.

Molecular techniques
Southern blotting and nucleic acid hybridization

A labelled DNA probe will bind to the specimen if it contains the specific sequence sought. The bound probe is detected by the activity of the label. This is a specific and rapid technique but less sensitive than methods involving amplification steps.

Nucleic acid amplification methods

Several nucleic acid amplification methods are used for diagnosis of infection. Each uses a slightly different method of amplifying pathogen target DNA or RNA until sufficient copies are available for detection. For example, in nucleic acid amplification testing (NAAT), pathogen DNA is separated into single strands and primers are designed to bind to target sequences then a polymerase catalyses synthesis of new DNA. A positive result can be obtained from as little as one copy of the target DNA. Automated systems and commercial kits have made these tests available in many laboratories. Real-time machines produce positive results quickly. Nucleic acid amplification techniques are valuable for the diagnosis of organisms that are difficult, slow or dangerous to grow, such as *Mycobacterium tuberculosis* and *Chlamydia trachomatis*. Methods can be used to detect antibiotic resistance genes providing a surrogate susceptibility result (e.g. detecting a *rpoB* gene mutation for rifampicin resistance in *M. tuberculosis*).

Typing

It is sometimes necessary to type organisms in order to follow their transmission in the hospital or community.

5 Antibacterial therapy

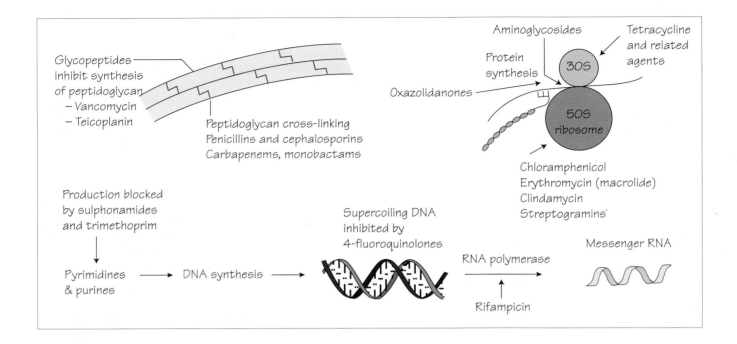

Principles of antibiotic therapy

Antibacterial chemotherapy depends on selective toxicity: the antibiotic interferes with the metabolism of the pathogen but not that of the host. It is best achieved by exploiting bacterial characteristics that are not present in human cells (e.g. unlike human cells, bacterial cells possess a cell wall). Inhibiting cell wall synthesis will inhibit the bacterium, but is unlikely to harm the host.

Appropriate antibiotic treatment is usually very effective and safe. While all antimicrobials have potential unwanted effects, serious ones are not frequent. Most antibiotics have a wide therapeutic index: the dose at which unwanted effects occur is very much higher than that which inhibits bacterial growth. An exception is the aminoglycosides, where the serum concentration must be carefully controlled (see below).

Choice of therapy

The choice of antibiotic depends on the site of infection, the susceptibilities of the likely infecting organisms, the severity of infection, any history of allergy, the likelihood of unwanted effects and lastly cost. Knowledge of the organism likely to infect a particular site and its antibiotic susceptibility profile should lead to a rational choice of therapy.

Site of infection Antibiotic penetration into tissues such as bone, joints and CSF varies. High levels of antimicrobial activity are difficult to achieve in abscesses and areas with poor blood supply. Low pH inhibits some antibiotic activity (e.g. aminoglycosides). The problem is magnified when the abscess lies within bone or in the CSF.

Organism Streptococcus pyogenes is invariably susceptible to penicillin, but others such as *Acinetobacter* and *Pseudomonas* are often multiply resistant, making antimicrobial choice difficult.

Likelihood of allergy/unwanted effects Many patients report that they are allergic to one or more antibiotic, most commonly to penicillins. An alternative therapy can usually be selected. When true allergy occurs it is usually a firm contraindication to the use of the inducing drug and often also to those in the same class. Renal impairment may be a contraindication to aminoglycosides. Age may be a contraindication to the use of cephalosporins which may increase the risk of super infection (e.g. with *C. difficile*).

Route of administration

The oral route of administration is commonly used, both in hospital and in community practice. Antibiotics may also be given topically for skin infections, *per rectum* (e.g. metronidazole for surgical prophylaxis), or vaginally as pessaries. Intravenous therapy is usually required in severe infections, such as septicaemia, to ensure adequate antibiotic concentrations. This route may also be chosen for patients unable to tolerate oral therapy, such as those with repeated vomiting. The palatability of paediatric formulations and the likelihood of patient compliance with frequent or complex regimens must also be considered.

Monitoring therapy

Antibiotic monitoring may be necessary either to ensure that adequate therapeutic levels have been achieved, or to reduce the risk of toxicity. This is especially important where the therapeutic range is close to the toxic range. Serum levels of both aminoglycosides and vancomycin are measured in blood samples taken just before and 1 h after intravenous or intramuscular dosage. Timed levels must be adjusted according to specific guidelines to ensure adequate antibacterial activity and reduce the risk of toxicity. For example, if the peak is high the dosage may be reduced; a high trough level can be lowered by taking medication less frequently. Levels taken using newer, once-daily regimes are interpreted using normograms and careful adherence to guidelines.

Serum concentrations are also helpful in the management of partially resistant organisms. If inhibition of an organism only occurs at high antibiotic concentrations, then it follows that it is important to maintain such levels in the circulation. When such an infection arises in a difficult site, e.g. *Pseudomonas* meningitis, antibiotic concentrations may be measured in the CSF.

Adverse events

Mild gastrointestinal upset is probably the most frequent side-effect of antibiotic therapy. Rarely, severe allergic reactions may lead to acute anaphylactic shock or serum sickness syndromes.

Gastrointestinal tract

Antibiotic activity can upset the balance of the normal flora within the gut: β-lactams are especially likely to do this, resulting in overgrowth of commensal organisms such as *Candida* spp. Alternatively, therapy may provoke diarrhoea or, more seriously, pseudomembranous colitis (see Chapter 18).

Skin

Cutaneous manifestations range from mild urticaria or maculopapular, erythematous eruptions to erythema multiforme and the life-threatening Stevens–Johnson syndrome. Most cutaneous reactions are mild and resolve after discontinuation of therapy.

Haemopoietic system

Patients receiving chloramphenicol or antifolate antibiotics may exhibit dose-dependent bone-marrow suppression. More seriously, aplastic anaemia may rarely complicate chloramphenicol therapy. High doses of β-lactam antibiotic may induce a granulocytopenia. Antibiotics are a rare cause of haemolytic anaemia. Many antibiotics cause a mild reversible thrombocytopenia or bone-marrow depression.

Renal system

Aminoglycosides may cause renal toxicity by damaging the cells of the proximal convoluted tubule. Elderly people, patients with pre-existing renal disease or those who are also receiving other drugs with renal toxicity are at higher risk. Tetracyclines may also be toxic to the kidneys.

Liver

Isoniazid and rifampicin may cause hepatitis: this is more common in patients with pre-existing liver disease. Other agents associated with hepatitis are tetracycline, erythromycin, pyrazinamide, ethionamide and, very rarely, ampicillin or fluoroquinolones. Cholestatic jaundice may follow tetracycline or high-dose fusidic acid therapy.

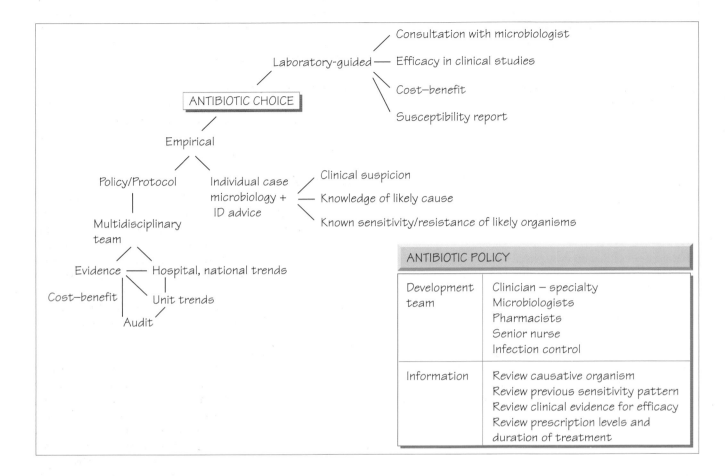

Beta-lactam antibiotics

Penicillins work by inhibiting peptidoglycan cross-linking. Natural penicillins have been modified to have penicillinase resistance or an extended antibacterial spectrum:

- natural penicillins (e.g. penicillin G, penicillin V)
- penicillinase-resistant penicillin (e.g. flucloxacillin)
- aminopenicillins (e.g. ampicillin-like agents)
- expanded-spectrum penicillins (e.g. piperacillin)
- penicillins and β-lactam inhibitors (e.g. amoxicillin and clavulanate).

Oral absorption varies: penicillin G is unstable in the presence of gastric acid and must be given intravenously, but penicillin V is stable and can be given orally. The aminopenicillins and flucloxacillin are also absorbed orally, while the remaining agents must be given intravenously.

Penicillins are rapidly secreted by the kidney and their half-life is very short. Probenecid competes for secretion and increases the half-life. Penicillins are distributed in extracellular fluid, not crossing the blood–brain barrier unless the meninges are inflamed.

Cephalosporins

Cephalosporins are closely related to penicillins; there are five classes.

1 Oral cephalosporins with a mainly Gram-positive spectrum.
2 Injectable agents (cefuroxime) active against Gram-positive organisms such as *Escherichia coli* and some species of *Proteus*.
3 Newer injectable cephalosporin agents (cefotaxime or ceftriaxone), active against most Gram-negative organisms and *Streptococcus* spp.
4 The fourth group has the same wide spectrum as the third group, but can be administered by the oral route.
5 Antipseudomonal, e.g. broad spectrum that additionally includes *Pseudomonas* (e.g. ceftazidime).

Monobactams

The monobactams are related to penicillins and cephalosporins by the presence of a β-lactam ring. They are broad spectrum, including anti-anaerobe activity. Imipenem and meropenem have antipseudomonal effects. They must be given intravenously.

Aminoglycosides

Aminoglycosides act by preventing translation of mRNA into protein. They are given parenterally and are limited to the extracellular fluid; they are excreted in the urine. Aminoglycosides are toxic to the kidney and eighth cranial nerve at amounts close to therapeutic levels, necessitating careful monitoring of serum concentrations.

Glycopeptides (vancomycin, teicoplanin)

The glycopeptides inhibit peptidoglycan cross-linking in Gram-positive organisms only. Bacterial resistance, previously uncommon, is now found in enterococci isolated in hospitals (glycopeptide-resistant enterococci—GRE) and some *Staphylococcus aureus*. They must be administered intravenously or intraperitoneally; they are not absorbed orally. The exception is the oral use of vancomycin to treat pseudomembranous colitis. Glycopeptides are distributed in the extracellular fluid, and do not cross the blood–brain barrier unless there is meningeal inflammation. They are excreted by the kidney. Daptomycin, a newer related agent with some promise, is very active against Gram-positive organisms with more rapid killing *in vitro*.

Quinolones

Quinolones kill by inhibiting bacterial DNA gyrase. The first quinolones did not attain high tissue levels and were used for urinary tract infection. Fluoroquinolones are more active against Gram-negative pathogens including *Pseudomonas*, and *Chlamydia*. They have been used for single-dose treatment of genital infections. Fluoroquinolones are well absorbed orally, are widely distributed and penetrate cells well. Newer agents (e.g. moxifloxacin) are more active against Gram-positive pathogens, including *Streptococcus pneumoniae* and *Mycobacterium tuberculosis*.

Macrolides (erythromycin, azithromycin and clarythromycin)

Macrolides bind to the 50S ribosome, interfering with protein synthesis, and are active against Gram-positive cocci, many anaerobes (but not *Bacteroides*), *Mycoplasma* and *Chlamydia*. Absorbed orally, they are distributed in the total body water, cross the placenta, are concentrated in macrophages, polymorphs and the liver and are excreted in the bile. Erythromycin may cause nausea. Newer macrolides have more favourable pharmacokinetics and toxicity profiles.

Streptogramins

Pristinomycin is a bactericidal semisynthetic streptogramin consisting of quinupristin and dalfapristin. It acts by preventing peptide bond formation, resulting in release of incomplete polypeptide chains from the donor site. It is active against a broad range of Gram-positive pathogens and some Gram-negatives, such as *Moraxella*, *Legionella*, *Neisseria meningitidis* and *Mycoplasma*. It is used mainly for the treatment of resistant Gram-positive infections (e.g. GRE and glycopeptide-intermediate *S. aureus*, GISA).

Oxazolidinones

The oxazolidinones inhibit protein synthesis at the 50S ribosomal subunit. They are most active against Gram-positive bacteria and are used mainly for the treatment of resistant Gram-positive infections.

Metronidazole

Metronidazole is active against all anaerobic organisms and acts by accepting electrons under anaerobic conditions and forming toxic metabolites that damage bacterial DNA. Metronidazole is also active against some species of protozoa, including *Giardia*, *Entamoeba histolytica* and *Trichomonas vaginalis*. It is absorbed orally and can be administered parenterally. Metronidazole is widely distributed throughout the tissues, crossing the blood–brain barrier and penetrating into abscesses. It is metabolized in the liver and excreted in the urine, and is well tolerated.

Tetracyclines

Tetracyclines interfere with protein synthesis by locking tRNA to the septal site of mRNA. They are active against many Gram-positive and some Gram-negative pathogens, *Chlamydia*, *Mycoplasma*, *Rickettsia* and treponemes. Doxycycline has useful activity against some protozoa including *Plasmodium* and *Entamoeba histolytica*. Absorbed orally, doxycycline has a long half-life, and adequate therapeutic levels may be obtained by a once-a-day dosage. The drug is distributed to many tissues including the lung, liver, kidney, brain and respiratory tract, and is concentrated in bile. Newer related agents such as tigecycline show promise in the treatment of resistant Gram-negative infections.

Sulphonamides and trimethoprim

Sulphonamides and trimethoprim act by inhibiting the synthesis of tetrahydrofolate. Trimethoprim and sulphonamide are now rarely used in the treatment of bacterial infections but have an important role in the management of *Pneumocystis jiroveci* and protozoan infections including malaria. Sulphonamides can be given intravenously and are well absorbed when given orally. They are distributed widely in the tissues and cross the blood–brain barrier. They are metabolized in the liver and excreted via the kidney.

7 Resistance to antibacterial agents

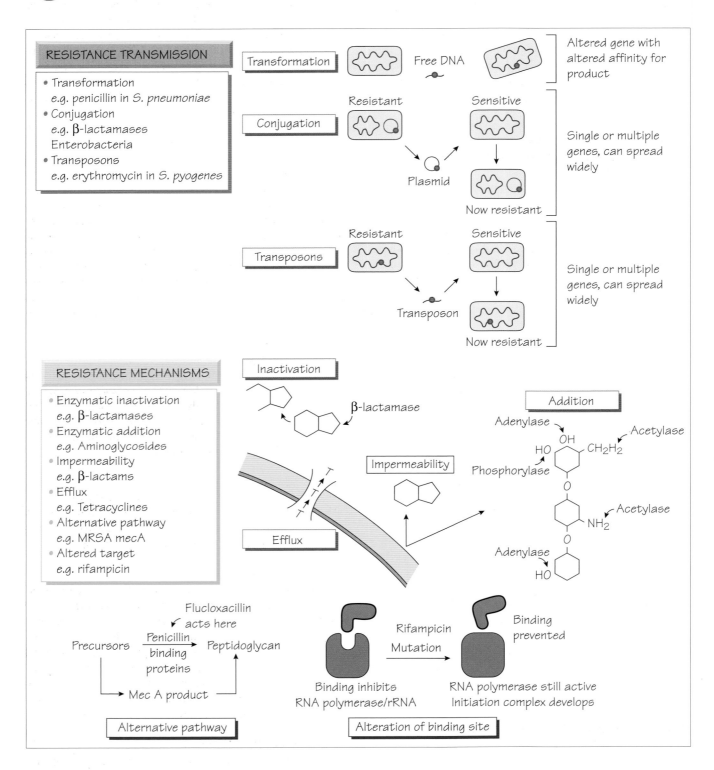

RESISTANCE TRANSMISSION

- Transformation
  e.g. penicillin in *S. pneumoniae*
- Conjugation
  e.g. β-lactamases
  Enterobacteria
- Transposons
  e.g. erythromycin in *S. pyogenes*

Transformation — Free DNA — Altered gene with altered affinity for product

Conjugation — Resistant — Sensitive — Plasmid — Now resistant — Single or multiple genes, can spread widely

Transposons — Resistant — Sensitive — Transposon — Now resistant — Single or multiple genes, can spread widely

RESISTANCE MECHANISMS

- Enzymatic inactivation
  e.g. β-lactamases
- Enzymatic addition
  e.g. Aminoglycosides
- Impermeability
  e.g. β-lactams
- Efflux
  e.g. Tetracyclines
- Alternative pathway
  e.g. MRSA mecA
- Altered target
  e.g. rifampicin

Inactivation — β-lactamase

Impermeability

Efflux

Addition
Adenylase — Acetylase
Phosphorylase
OH CH_2H_2
HO
Acetylase
NH_2
Adenylase
HO

Precursors — Penicillin binding proteins → Peptidoglycan
Flucloxacillin acts here
Mec A product
Alternative pathway

Rifampicin Mutation
Binding inhibits RNA polymerase/rRNA
Binding prevented
RNA polymerase still active
Initiation complex develops
Alteration of binding site

'Resistance' occurs when a previously susceptible organism is no longer inhibited by an antibiotic at levels that can be safely achieved clinically. This happens because the bacterial gene pool changes, facilitated by its rapid cell division and haploid genome. Organisms may transfer genetic material within and between species. Bacteria do not have a deliberate policy to develop 'resistance genes' or 'virulence factors' to advance their species: they play the genetic lottery. Antibiotic use allows the survival and replication of organisms that have accidentally developed mechanisms to avoid destruction.

Transmission of resistance determinants between bacteria

Transformation

Many bacterial species can take up naked DNA and incorporate it into their genome: this is called transformation. For example, *Streptococcus pneumoniae* takes up part of penicillin-binding protein genes from closely related species. The altered gene produces a penicillin-binding protein which binds penicillin less avidly and so is not inhibited by penicillin to the same extent. The organism is still able to synthesize peptidoglycan and maintain its cell wall in the presence of penicillin. Resistance to penicillin by *Neisseria gonorrhoeae* also develops in the same way.

Conjugation

Plasmids are circular DNA structures found in the cytoplasm. Multiple copies may be present and, following cell division, they are found in the cytoplasm of each daughter cell. Many genes are carried on plasmids, including those coding for metabolic enzymes, virulence determinants and antibiotic resistance. The process of conjugation occurs when plasmids are passed from one bacterium to another. In this way 'resistance genes' can spread rapidly in populations of bacterial species that share the same environment (e.g. within the intestine). Combined with antibiotic selective pressure (e.g. in hospitals) multiresistant populations may develop.

Transposons and integrons

Transposons and integrons are moveable genetic elements able to encode transposition. They can move between the chromosome and plasmids, and between bacteria. Many functions, including antibiotic resistance, can be encoded on a transposon. Resistance to methicillin among *Staphylococcus aureus* and that of *Neisseria gonorrhoeae* to tetracycline probably entered the species by this route. Integrons are important in transmission of multiple drug resistance in Gram-negative pathogens. Resistance genes can also be mobilized by bacteriophages.

Mechanisms of resistance

Antibiotic modification

Enzyme inactivation

One of the most common resistance mechanisms occurs when the organism spontaneously produces an enzyme that degrades the antibiotic. Many strains of *Staphylococcus aureus* produce an extracellular enzyme, β-lactamase, which breaks open the β-lactam ring of penicillin, inactivating it. Many other organisms are capable of expressing enzymes that degrade penicillins and cephalosporins. These include *Escherichia coli*, *Haemophilus influenzae* and *Pseudomonas* spp. Often the genes that code for these enzymes can be found on mobile genetic elements (transposons) and can be transmitted between organisms of different species. The spread of different types of extended-spectrum β-lactamases (ESBLs) such as CTXm and AmpC among enterobacteriaceae are resulting in resistance to broad-spectrum penicillins and cephalosporins in organisms causing hospital-associated infections. Spread to the community is likely.

Enzyme addition

Bacteria may express enzymes that add a chemical group to the antibiotic, inhibiting its activity. Bacteria become resistant to aminoglycosides by expressing enzymes that inactivate the antibiotic by adding an acetyl, amino or adenosine group to the antibiotic molecule. The different members of the aminoglycoside family differ in their susceptibility to this modification, amikacin being the least susceptible. Aminoglycoside resistance enzymes are possessed by Gram-positive organisms, such as *Staphylococcus aureus*, and Gram-negative organisms, such as *Pseudomonas* spp.

Impermeability

Some bacteria are naturally resistant to antibiotics because their cell envelope is impermeable to particular antibiotics. Gram-negative organisms, especially *Pseudomonas* spp., are impermeable to some β-lactam antibiotics. Aminoglycosides enter bacteria by an oxygen-dependent transport mechanism and so have little effect against anaerobic organisms.

Efflux mechanisms

Bacteria, for example *E. coli*, may become resistant to tetracyclines by the acquisition of an inner membrane protein which actively pumps the antibiotic out of the cell. Streptococci may become resistant to macrolides using an efflux pump.

Alternative pathway

Another common bacterial mechanism is the development of an alternative pathway to circumvent the metabolic block imposed by the antibiotic. *Staphylococcus aureus* becomes resistant to methicillin or flucloxacillin when it acquires the gene *mecA*. This codes an alternative penicillin-binding protein (PBP2′) which is not inhibited by methicillin. Although the composition of the cell wall is altered, the organism is still able to multiply. Similar alterations to the penicillin-binding proteins of *Streptococcus pneumoniae* are responsible for resistance in this organism.

Alteration of the target site

Rifampicin acts by inhibiting the β-subunit of RNA polymerase. Resistance develops when the RNA polymerase gene is altered by point mutations, insertions or deletions. The new RNA polymerase is not inhibited by rifampicin and resistance occurs.

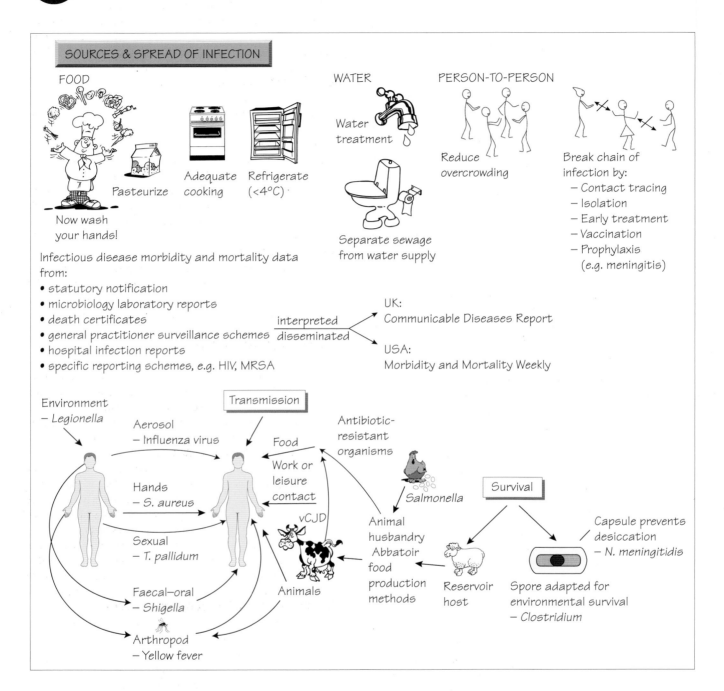

Sources of infection

Clinical infection can be caused either by organisms that originate from the host's normal flora (endogenous infections) or by organisms that are acquired from another source (exogenous infections).

Endogenous organisms of the normal flora only invade under certain circumstances, such as when the bowel is perforated. Enterobacteriaceae and non-sporing anaerobes such as *Bacteroides fragilis* cause intraperitoneal abscesses. Inhalation of stomach contents or oropharyngeal secretions containing mixed facultative and anaerobic flora may cause inhalation pneumonia or lung abscess. *Staphylococcus aureus* normally carried innocently in the anterior nares may cause a wound infection if inoculated into a surgical wound from the patient's own nasal secretions. Neutropenic sepsis results when gut bacteria escape the intrinsic defences of the gut and normal neutrophil function during the neutropenic phase of chemotherapy for leukaemia, to cause bacteraemia. Changes in the host alter the risk of disease: surgery and intravenous cannulation favour invasion with organisms of the normal flora, and immunosuppressive therapy makes patients susceptible to opportunists of low virulence.

Exogenous sources are variable. Animal pathogens may spread to humans by contact or in food; these infections are called zoonoses. Humans can become infected from organisms in the inanimate environment, such as *Legionella* or *Clostridium*.

Alteration of the environment changes the risk of disease. Zoonoses are encouraged by intensive farming methods. Feeding ruminant offal to cattle resulted in an epidemic of bovine spongiform encephalopathy (BSE) which then spread to humans as variant Creutzfeldt–Jakob disease (vCJD). Battery farming encourages the spread of *Salmonella* throughout poultry flocks, and mechanized food production techniques increase the likelihood of cross-contamination. Introduction of good husbandry techniques to interrupt contamination in food plants helps counter this. Poorly maintained air-conditioning cooling towers can be a source of *Legionella pneumophila*.

Some microorganisms have developed complex life cycles to facilitate transmission and survival. Organisms excreted in faeces spread to other hosts by ingestion: the faecal–oral route. Others have a life-cycle stage inside an insect vector which transmits the disease by biting. Humans can become infected as an accidental host when they substitute for an animal in a life-cycle (e.g. hydatid disease; see Chapter 52).

Survival and transmission

Organisms must survive in the environment. Spores are small structures with a tough coat and a low metabolic rate which enable bacteria to survive for many years. Helminth eggs have a tough shell adapted for survival in the environment. Transmission is favoured when an organism is able to survive long term in a host, which then acts as a reservoir of infection.

Microorganisms are propelled out of the nose and mouth in a sneeze and can remain suspended in the air on droplet nuclei (5 μm). Infection may occur when these are inhaled by another person and are carried to the alveoli. Respiratory infections such as influenza are transmitted this way, as are others which affect other organs (e.g. *Neisseria meningitidis*).

Food and water contain pathogens that may infect the intestinal tract (e.g. *Salmonella*). Toxoplasmosis and cysticercosis, which affect other organs, can be transmitted by this route.

Leptospira, *Treponema* and *Schistosoma* have evolved specific mechanisms enabling them to invade intact skin. Injections and blood transfusions bypass the skin, allowing the transmission of HIV. Skin organisms (e.g. *Staphylococcus epidermidis*) can invade the body via indwelling venous cannulae. Insects that feed on blood may transmit pathogens: anophelene mosquitoes transmit malaria.

Sexual intercourse is a route of spread for organisms with poor ability to survive outside the body, such as *Neisseria gonorrhoeae* or *Treponema pallidum*. Transmission is enhanced by genital ulceration.

Social and environmental factors

Improvements in social and environmental conditions can reduce the burden of infectious disease. For example, improved sanitation reduces the risk of diarrhoeal diseases and better housing reduces the spread of tuberculosis. Better nutrition means that the population is less susceptible to disease.

Paradoxically, the morbidity from some infectious diseases may rise as living conditions improve. This occurs when the complication rate is higher in adults than in children, e.g. in paralytic poliomyelitis (see Chapter 34) or varicella zoster infection.

Health education

There are many effective infection-related health education programmes covering safe sex, needle exchange, advice to pregnant women, guidance on food hygiene and advice to travellers.

Food safety

Food safety legislation has been harmonized across the European Union. The law is enforced in food premises by environmental health officers (EHOs) and by officials of the Ministry of Agriculture, Fisheries and Food (DEFRA) on farms. Milk pasteurization reduces the risk of infection with *Mycobacterium bovis* and *Campylobacter* spp.

Vector control

Vector control is important where arthropods transmit infections. Travellers to the tropics can reduce the risk of infection by taking measures to avoid insect bites. Attempts to control insect populations using pesticides have usually been unsuccessful because of insecticide resistance.

Chemoprophylaxis

Chemoprophylaxis is used for control of some serious infections, such as diphtheria and meningococcal disease. It aims to eliminate carriage of pathogens to prevent further spread and cases. For example, rifampicin or ciprofloxacin is given to meningococcal contacts. Isoniazid is given to patients at risk of TB reactivation should they become immunosuppressed.

Outbreak investigation

Basic epidemiological information is collected, e.g. onset of symptoms, age, sex, place of residence, and a detailed food history.

A hypothesis of causation is tested by a case–control or cohort study: exposure histories are sought from cases and healthy controls. The relative risk of exposure is calculated for each group. Case–control studies are suited to investigation of outbreaks of uncommon infections, such as botulism, and infections with a high attack rate, such as food poisoning. Cohort studies compare the disease outcome between those exposed and not exposed.

The role of national agencies

Most countries have a national system to control communicable diseases. It has four main functions:
1 surveillance of communicable diseases
2 investigation of outbreaks
3 surveillance of immunization programmes
4 epidemiology research and training.

Close collaboration between food and agriculture control agencies and the human infection control agency is required for zoonotic infections. In some countries governments have combined these functions in a single agency, for example the Health Protection Agency in the UK and the Centers for Disease Control in the US. These national agencies are undergoing review in the light of the bioterrorism threat, with the increased need for early detection of unusual infections that may be indicators of a deliberate release of a biological agent.

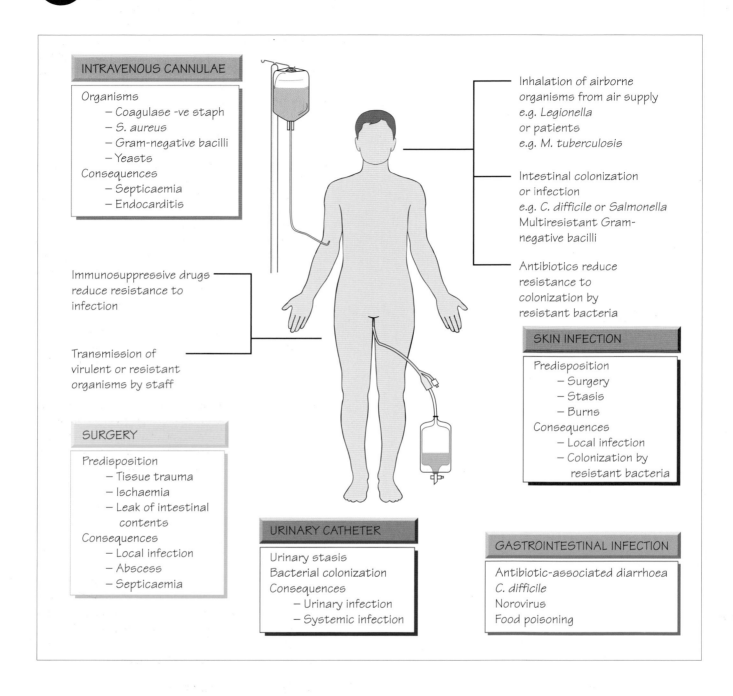

INTRAVENOUS CANNULAE

Organisms
- Coagulase -ve staph
- S. aureus
- Gram-negative bacilli
- Yeasts
Consequences
- Septicaemia
- Endocarditis

Inhalation of airborne
organisms from air supply
e.g. Legionella
or patients
e.g. M. tuberculosis

Intestinal colonization
or infection
e.g. C. difficile or Salmonella
Multiresistant Gram-
negative bacilli

Antibiotics reduce
resistance to
colonization by
resistant bacteria

Immunosuppressive drugs
reduce resistance to
infection

Transmission of
virulent or resistant
organisms by staff

SKIN INFECTION

Predisposition
- Surgery
- Stasis
- Burns
Consequences
- Local infection
- Colonization by
  resistant bacteria

SURGERY

Predisposition
- Tissue trauma
- Ischaemia
- Leak of intestinal
  contents
Consequences
- Local infection
- Abscess
- Septicaemia

URINARY CATHETER

Urinary stasis
Bacterial colonization
Consequences
- Urinary infection
- Systemic infection

GASTROINTESTINAL INFECTION

Antibiotic-associated diarrhoea
C. difficile
Norovirus
Food poisoning

Hospital-acquired infection is infection which was not present or incubating at the time of admission. It is very common (occurring in up to 25% of patients admitted). The most frequent types of infection are urinary tract, respiratory, wound, skin and soft-tissue infection, and septicaemia (often associated with vascular access).

The environment
The potential for person-to-person transmission of organisms within hospitals is enormous.

Food supply
Food is usually prepared centrally in the hospital kitchens: patients are at risk of food-borne infection if hygiene standards fall. Antibiotic-resistant organisms can be transmitted by this route.

Air supply
Pathogens, e.g. multidrug-resistant tuberculosis or respiratory viruses may be transmitted via theatre air supply and air-conditioning systems. Badly maintained air-conditioning systems may be a source of *Legionella*.

Fomites

Any inanimate object may be contaminated with organisms and act as a vehicle (fomite) for transmission.

Water supply

The water supply in the hospital is a complex system, supplying water to wash-hand basins and showers, central heating and air-conditioning. Additionally, superheated steam at pressure is required for autoclaves. *Legionella* spp. may colonize the system in redundant areas of pipework. Cooling-tower systems are a particular source of infection, allowing transmission via the air-conditioning system. To reduce this risk, hot-water supplies should be maintained at a temperature above 45°C and cold-water supplies below 20°C.

The host

Hospital patients are susceptible to infection as a result of underlying illness or treatment, for example patients with leukaemia or taking cytotoxic chemotherapy. Age and immobility may predispose to infection; ischaemia may make tissues more susceptible to bacterial invasion.

Medical activities
Intravenous access

This is the most frequent source of healthcare-associated bacteraemia. The risk of infection from any intravenous device increases with the length of time it is in position. Having broken the skin's integrity, it provides a route for invasion by skin organisms such as *Staphylococcus aureus*, *S. epidermidis* and *Corynebacterium jeikeium*. Signs of inflammation at the puncture site may be the first evidence of infection. Cannula-related infection can be complicated by septicaemia, endocarditis and metastatic infections (e.g. osteomyelitis). Aseptic technique at insertion will reduce the risk of sepsis as will the choice of device, i.e. those without side ports and dead spaces. Maintaining adequate dressing and ensuring good staff hygiene when working with the device are equally important. The cannula site should be regularly inspected and this is particularly important in unconscious patients. Peripheral lines should be re-sited every 48 h; central and tunnelled lines should be changed when there is evidence of infection.

Urinary catheters

Indwelling urinary catheters provide a route for ascending infection into the bladder. Risks can be minimized by aseptic technique when the catheter is inserted and handled.

Surgery

Surgical patients often have other health problems that are unrelated to their surgical complaint (e.g. asthma or diabetes mellitus), and these may predispose them to infection. Surgery is traumatic and carries a risk of infection, e.g. especially wound infections. In addition, there are the potential complications of the procedure itself, such as postoperative ischaemia, that contribute further risks. The length and complexity of the operation influence the risk of infection, as does the skill of the surgeon: the less tissue damage that occurs at the time of operation the lower the risk of infection. The preoperative period should be short to reduce the risk of acquiring resistant hospital organisms. Elective surgery should be postponed for patients with active infection (e.g. chest infections).

To minimize the risk of infection during an operation, theatres are supplied with a filtered air supply. Staff movement during procedures should be limited to reduce air disturbance. Changing clothing reduces transmission of organisms from the wards. Impervious materials reduce contamination from the skin of the surgical team but are uncomfortable to wear. Some hospitals provide ventilated air-conditioned suits for surgical teams performing prosthetic joint surgery.

Antibiotic prophylaxis can reduce the risk of postoperative infection. Those chosen should be bactericidal, and penetrate to the required site at sufficient concentrations to be active against organisms normally implicated in infection. There is no evidence that continuing prophylaxis beyond 48 h is beneficial.

'Clean' operations involve only the skin or a normally sterile structure, such as a joint. Such operations do not need antibiotics unless a prosthetic device is inserted when antibiotics active against staphylococci should be given.

'Contaminated' operations are those in which a viscus that contains a normal flora is breached. Appropriate antibiotics might be metronidazole with a second-generation cephalosporin for large-bowel surgery; cephalosporin alone is satisfactory in upper gastrointestinal tract or biliary tract surgery where anaerobes are rarely implicated.

'Infected' operations are those in which surgery is required to deal with an already infected situation, such as drainage of an abscess or repair of a perforated diverticulum. Systemic antibiotics directed against the likely infecting organisms should be prescribed.

Intubation gives organisms access to the lower respiratory system. Postoperative pain, immobility and the effects of anaesthesia may predispose to pneumonia by reducing coughing. Respiratory infections with resistant Gram-negative organisms originating from the hospital environment may occur.

Quality improvement and healthcare-associated infection targets

Improvement targets with mandatory reporting of key measures have focused attention on healthcare-associated infections, e.g. MRSA and *Clostridium difficile*.

10 Control of infection in hospital

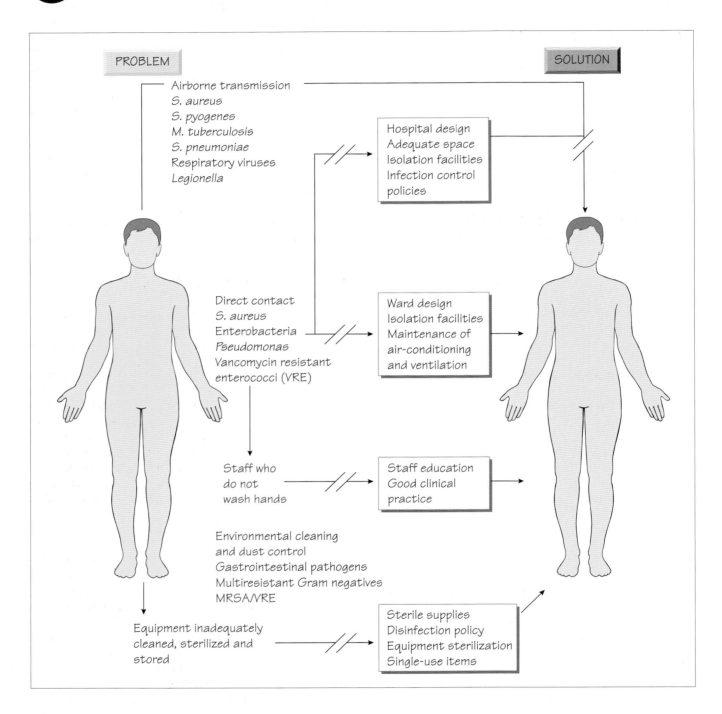

PROBLEM

SOLUTION

Airborne transmission
S. aureus
S. pyogenes
M. tuberculosis
S. pneumoniae
Respiratory viruses
Legionella

Hospital design
Adequate space
Isolation facilities
Infection control
policies

Direct contact
S. aureus
Enterobacteria
Pseudomonas
Vancomycin resistant
enterococci (VRE)

Ward design
Isolation facilities
Maintenance of
air-conditioning
and ventilation

Staff who
do not
wash hands

Staff education
Good clinical
practice

Environmental cleaning
and dust control
Gastrointestinal pathogens
Multiresistant Gram negatives
MRSA/VRE

Equipment inadequately
cleaned, sterilized and
stored

Sterile supplies
Disinfection policy
Equipment sterilization
Single-use items

Every hospital should have procedures to ensure that infection is not transmitted within its environment. Together, these form the infection control policy that, if it is to be successful, must have the support of the entire hospital staff. The control of infection team, consisting of a consultant microbiologist or infectious diseases specialist, and specialist nurses, promotes the policy. Involvement of the hospital management at the highest level is essential for success.

The team will arrange enhanced surveillance of particular organisms, e.g. methicillin-resistant *Staphylococcus aureus*

(MRSA). It also has a role in all hospital planning, both physical (e.g. alterations to buildings) and functional (e.g. new clinical services).

Good clinical practice
Infected individuals should be separated from non-infected. Sources, infected or colonized (carriers), must be identified by appropriate screening measures, e.g. routine surveillance specimens from both patients and staff. Infected patients should be isolated (source isolation) and practical measures taken to interrupt chains of transmission. Patients who are especially susceptible to infection require protective isolation. Isolation is often difficult to maintain when staff do not adhere to agreed practice. This may be compounded when simple measures such as hand-washing are neglected as a result of work pressures.

Wound and enteric isolation
Patients are nursed in a side-room that contains a wash-hand basin and separate toilet facility. Disposable plastic aprons and gloves are used while handling the patient or performing clinical and personal hygiene procedures. The gloves and apron are then discarded and hands washed using liquid soap and disposable towels. Appropriate use of disinfectant helps reduce environmental contamination (e.g. chlorine-containing agents for *Clostridium difficile*; see Chapter 18).

Respiratory isolation
In addition to the precautions listed above, hospital staff should wear a facemask when in the room. If the patient is transferred to another department of the hospital, the patient should wear a facemask. Stricter respiratory isolation methods are necessary to control transmission of the organisms associated with multidrug-resistant tuberculosis (MDRTB) and severe acute respiratory syndrome (SARS). This requires the use of negative-pressure rooms and effective masks (dust mist masks or personal respirators). Such precautions are especially essential during procedures that are likely to generate aerosols (e.g. bronchoalveolar lavage).

Strict isolation
This form of isolation is designed to prevent the transmission of infections such as viral haemorrhagic fevers. An enclosed isolation unit prevents aerosol transmission of the organism by its enclosed air system and negative pressure, together with strict decontamination procedures.

Protective isolation
Protective isolation is required for patients who are highly susceptible to infection, such as neutropenic patients. Protection includes single-room isolation, provision of filtered air and measures to control the risk from organisms in food, such as resistant Gram-negative organisms in vegetables or *Listeria* in soft cheeses.

Typing
Typing determines if organisms are identical and whether cross-infection has occurred. Chosen techniques should be simple to perform and reproducible, giving similar results when used in another laboratory.
- **Simple laboratory typing**: using phenotypic markers.
- **Serological typing**: suitable for testing *Shigella flexneri* or the salmonellas.
- **Phage typing**: bacteriophages lyse the bacteria they infect. This phenomenon is used in phage typing, e.g. for staphylococci and some species of *Salmonella*.
- **Colicine typing**: some bacteria produce protein antibiotics such as colicines that inhibit closely related bacteria. This can be used to type *Shigella* and *Pseudomonas*.
- **Molecular typing**: restriction endonuclease enzymes are used to digest genomic or plasmid DNA or ribosomal RNA giving a characteristic pattern. Identical organisms will have identical band patterns. Nucleic acid sequence-based methods are emerging such as multilocus sequence typing (MLST) where a portion of seven housekeeping genes are sequenced and the results compared. Molecular typing is increasingly replacing other methods.

Sterilization and disinfection
Sterilization
Sterilization inactivates all infectious organisms and is achieved by autoclaving or irradiation. In the autoclave items such as surgical instruments, are heated with superheated pressurized steam to inactivate any contaminating infectious material. Delicate instruments can be sterilized at low pressures and temperatures in specialized autoclaves that deliver steam together with formaldehyde. Perishable materials, such as plastic cannulae, syringes or prosthetic devices, are sterilized using γ-irradiation during commercial manufacture. Aldehydes (e.g. glutaraldehyde and formaldehyde) are capable of sterilizing instruments if they are adequately cleaned first and the equipment is immersed for a sufficient length of time. Compounds such as chlorine dioxide are replacing glutaraldehyde to reduce toxicity to human operators.

Disinfection
This is the process of reducing the number of infectious particles. Simple washing with soaps or detergents is the most important component in disinfection. Disinfectants are chemicals that kill or inhibit microbes. They are used where it would be impossible to achieve sterile conditions (e.g. skin preparation before surgery), or after spillage of biological fluid (urine, blood or faeces) over an inanimate surface. Hypochlorite compounds (sodium hypochlorite, bleach), which are most active against viruses, are also useful after spillage, but are corrosive to metals. Halogen compounds, such as iodine, are active against bacteria, including spore-bearing organisms, but are relatively slow acting. They are used in disinfection of skin. Phenolic disinfectants are highly active against bacteria and are used to disinfect contaminated surfaces in the hospital and in bacteriology laboratories. Alcohol (70%) acts rapidly against bacteria, fungi and viruses and is useful in disinfecting skin preoperatively. Chlorhexidine is active against bacteria, especially staphylococci; it is also used for disinfection of the skin.

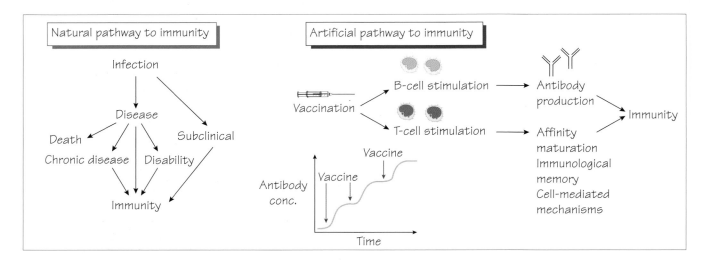

Immunization

Many infectious diseases are controlled by vaccination (e.g. polio and diphtheria). Smallpox has been completely eradicated by this means. Immunization may be achieved passively by administration of an immunoglobulin preparation, or actively by vaccination.

Immunoglobulins

Immunoglobulins provide short-term protection against certain infections and are useful in the management of immune disorders. Human immunoglobulin is prepared from pooled plasma and contains antibodies to viruses that are prevalent in the general population. Specific immunoglobulins, prepared from hyperimmune donors, are available for postexposure management, for example for hepatitis B, varicella zoster and tetanus. Tick-borne encephalitis immunoglobulin is available in countries where the disease is endemic.

Vaccination

The purpose of vaccination is to produce immunity in subjects without the complications of natural infection. Vaccines are derived from whole viruses and bacteria, or their antigenic components (acellular).

Live vaccines consist of strains where pathogenicity has been removed (attenuated), for example mumps, measles, rubella, polio and yellow fever vaccines. They usually produce durable immunity after a single dose. Live vaccines can cause disease in immunocompromised patients and are avoided in pregnancy because of the risk of fetal infection.

Non-replicating vaccines contain either inactivated whole organisms (e.g. pertussis) or antigenic components (e.g. capsular polysaccharide of *Streptococcus pneumoniae*). The toxins of tetanus and diphtheria are inactivated to produce toxoids that do not cause symptoms but are fully immunogenic – that is, they stimulate a protective antibody response. The immunogenicity of some acellular vaccines can be increased by conjugation with

proteins (e.g. *Haemophilus influenzae* type b). Genetic engineering is used for acellular vaccine production (hepatitis B). These vaccines are safe in immunocompromised patients because they are unable to replicate. Multiple doses may be required for optimum immunogenicity.

An immunization programme may have different aims: eradication of the pathogen (e.g. smallpox) resulting in total absence of the organism in humans, animals and the environment (only achievable where there is no non-human reservoir of the infection); elimination where the disease has largely disappeared but the organism remains in humans, animal hosts or the environment (e.g. clinical diphtheria and tetanus are almost prevented as is paralytic polio, but the organisms remain prevalent); or containment where the burden of disease or transmission is interrupted.

Universal immunization is adopted for most childhood infections, and selective programmes adopted for those at risk from disease (e.g. hepatitis B in healthcare workers). Minor differences occur in national immunization programmes across continents. These are updated regularly to take account of changing disease epidemiology and vaccine availability.

Vaccine safety

All clinical interventions carry some degree of risk. In determining vaccine policy the risks from natural disease must significantly outweigh the risks of vaccination. The safety of any vaccine is of paramount importance. Extensive trials are carried out to evaluate safety and strict 'good manufacturing practice' legislation governs the productions of vaccine. When a vaccine is given, records are kept of the circumstances of administration and vaccine lot number. A number of highly publicized false vaccine scares have undermined health professional and public confidence in a number of childhood vaccines. This in turn has reduced uptake and resulted in deaths and lifelong morbidity as a consequence of re-emergence of diseases such as whooping cough and measles.

Up-to-date information about immunization can be obtained in the UK from the Health Protection Agency (http://www.hpa.org.uk/infections/topics_az/vaccination/vac_guidelines.htm); in the US from the Infectious Diseases Society of America and the Centers for Disease Control and Prevention (http://www.idsociety.org/ and http://www.cdc.gov/); in the Pacific from http://immunise.health.gov.au/publications.htm; and from the World Health Organization (http://whqlibdoc.who.int/publications/2005/9241580364_chap6.pdf). This includes advice for travellers and about immunization schedules.

Current optimum vaccination schedule in developed countries are shown.

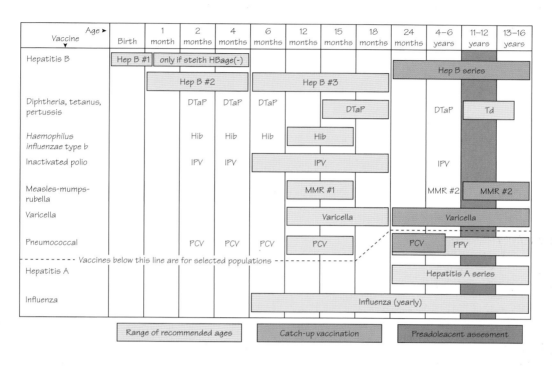

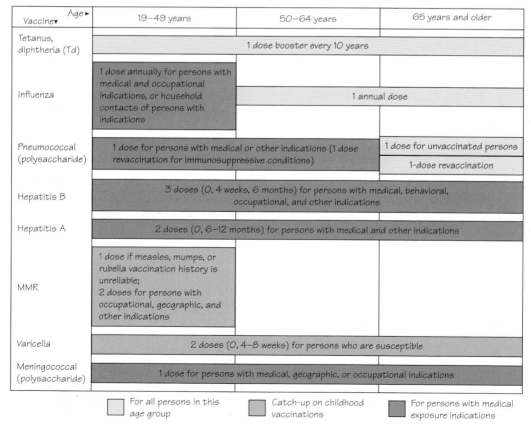

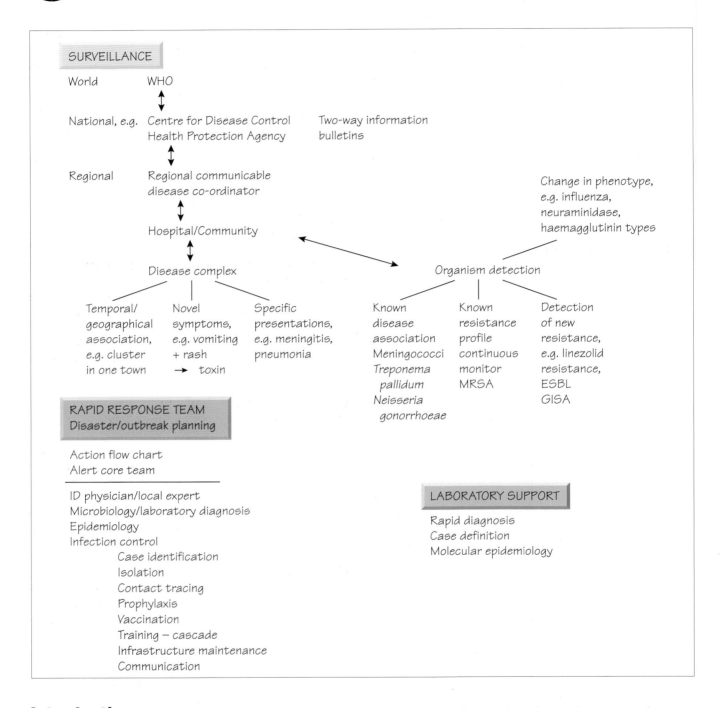

SURVEILLANCE

World — WHO

National, e.g. — Centre for Disease Control / Health Protection Agency — Two-way information bulletins

Regional — Regional communicable disease co-ordinator

Change in phenotype, e.g. influenza, neuraminidase, haemagglutinin types

Hospital/Community

Disease complex

| Temporal/ geographical association, e.g. cluster in one town | Novel symptoms, e.g. vomiting + rash → toxin | Specific presentations, e.g. meningitis, pneumonia |

Organism detection

| Known disease association Meningococci Treponema pallidum Neisseria gonorrhoeae | Known resistance profile continuous monitor MRSA | Detection of new resistance, e.g. linezolid resistance, ESBL GISA |

RAPID RESPONSE TEAM
Disaster/outbreak planning

Action flow chart
Alert core team

ID physician/local expert
Microbiology/laboratory diagnosis
Epidemiology
Infection control
    Case identification
    Isolation
    Contact tracing
    Prophylaxis
    Vaccination
    Training – cascade
    Infrastructure maintenance
    Communication

LABORATORY SUPPORT

Rapid diagnosis
Case definition
Molecular epidemiology

Introduction

Infectious diseases undergo constant change. The incidence of any infection rises and falls with changes in the immunity of the host population and through changes in the virulence of the pathogen. Many infectious diseases have a characteristic epidemic pattern. For example, meningococcal infection is normally endemic in a population, with epidemic peaks every 10–12 years. In the 'meningitis belt' of Africa the incidence of disease may rise to 1000 per 100 000 of the population.

Endemic infections are those that are always present in a community even though the number of cases may vary over time. Superimposed on this are the epidemics where the numbers of cases occurring exceed that which would normally be expected. The words 'outbreak' and 'epidemic' are often used interchangeably, although an outbreak is better thought of as a localized epidemic. A pandemic is an increase in the number of cases above that expected, spreading throughout the world. Examples of pandemics include the international spread of influenza after the organism undergoes an antigenic shift (see Chapter 33) and the spread of bubonic plague and cholera.

The speed at which an infection spreads through a community depends on the mechanism of transmission (see Chapter 8). For example, organisms that spread by the respiratory route can spread more rapidly than organisms that spread by the sexual route. The infectivity of a pathogen is defined by the speed at which an organism spreads in a community. Measles, for example, is highly infectious, whereas mumps is less so. It can be quantified using the 'intrinsic reproductive number' which is the average number of secondary cases arising from a single case in a totally susceptible population. The number of cases continues to rise and the number of susceptible individuals falls because of death or development of immunity. As the proportion of susceptible individuals falls, the number of new cases (incidence) will decline. Mathematical models can be used to predict the outcome of an epidemic and to indicate control methods.

Emerging infections

It has been recognized that those involved in the management of infectious disease must be alert to the threats posed by emerging infections. This is important because 'new' diseases may not be immediately recognized and thus go undiagnosed. Emerging diseases fall into three broad categories: new pathogens; invasion of a new territory by a pathogen; and re-emergence of a previously uncommon condition.

New pathogens

Several of the most important emerging infections have arisen because they are genuinely new infections. The most important example of this is the human immunodeficiency virus (HIV). This is highly homologous to simian immunodeficiency virus (SIV), and was thought to have jumped species from chimpanzees in central Africa just over 50 years ago, becoming adapted to its new human host. The numbers of cases have built up gradually and have spread from central Africa throughout the world (see Chapter 44). Severe acute respiratory syndrome (SARS) was caused by a novel coronavirus jumping species in southern China. Medicine, or changes in the environment, can create the conditions where species-hopping can occur. For example, there are concerns that xenotransplantation may provide the opportunity for animal viruses to infect immunocompromised transplant patients and spread to others from them. Destruction and farming of South American rain forests brought humans into contact with *Trypanosoma cruzi*, leading to human infections.

New territory

Climatic change or changes in population centres may permit an organism to invade new territory. For example, infection with West Nile virus is spreading in the US at present, entering some states for the first time. Climate change may allow organisms that require high ambient temperatures to survive as global temperatures rise.

Re-emergence of previously uncommon infections

In most industrialized countries tuberculosis has become uncommon. Through a combination of migration from countries of high endemicity, the HIV epidemic and neglect of public health precautions, the number of cases has risen rapidly. After the collapse of the Soviet Union diphtheria, which had previously been rare, re-emerged due to failings in vaccination programmes, emphasizing that many of the conquered infectious diseases await their opportunity to re-emerge. Multidrug-resistant bacteria may provide the opportunity for other infections to re-emerge.

Changes in vaccination uptake may allow previously controlled infectious disease to re-emerge and affect different age groups, e.g. whooping cough and measles in the UK.

Changes in agriculture and food production

The spread of organisms in livestock herds or flocks such as strains of *Salmonella enteritidis* spp. may provide new opportunities for contamination of products in the food chain. The widespread use of preprepared food has increased the risk of listeriosis in industrialized countries and has resulted in changes in food storage regulations.

Changes in the virulence or transmissibility of a pathogen

Some pathogens may emerge in particular niches and replace organisms of different or the same species because they are more virulent or more easily transmitted. *Clostridium difficile* 027 is a hypertoxin producer that has caused outbreaks in Canada, the US and the UK. *Acinetobacter baumanni* has caused an outbreak of infection in London hospitals. Epidemic MRSA (EMRSA) 15 and 16 have caused outbreaks in many settings and are now the predominant strains of methicillin-resistant *S. aureus* worldwide.

Bioterrorism

Increasing political uncertainty throughout the world has been coupled with increased terrorist activity. This has led to the possibility of groups of individuals using microbes as biological weapons. One bioterrorism incident occurred in the US using anthrax, resulting in four deaths and the need for the creation of extensive control measures. Other possible agents include smallpox, tularaemia, plague and a number of viral haemorrhagic fevers. Healthcare professionals require knowledge of unusual infections and their presentations, and need to be able to communicate concerns to agencies so that changing patterns can be identified quickly (see Chapter 8).

13 Staphylococcus

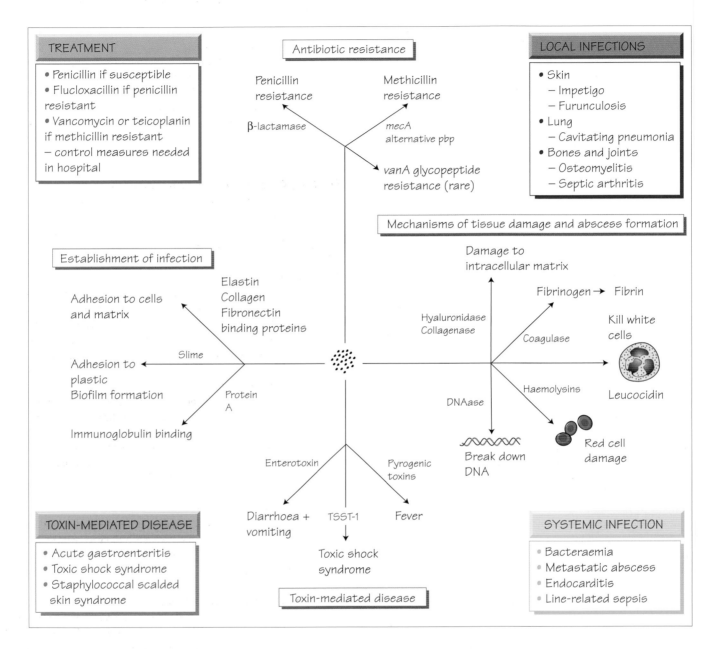

TREATMENT
- Penicillin if susceptible
- Flucloxacillin if penicillin resistant
- Vancomycin or teicoplanin if methicillin resistant
  – control measures needed in hospital

Antibiotic resistance

Penicillin resistance β-lactamase

Methicillin resistance mecA alternative pbp

vanA glycopeptide resistance (rare)

LOCAL INFECTIONS
- Skin
  – Impetigo
  – Furunculosis
- Lung
  – Cavitating pneumonia
- Bones and joints
  – Osteomyelitis
  – Septic arthritis

Mechanisms of tissue damage and abscess formation

Damage to intracellular matrix

Establishment of infection

Adhesion to cells and matrix

Elastin Collagen Fibronectin binding proteins

Slime

Adhesion to plastic Biofilm formation

Protein A

Immunoglobulin binding

Hyaluronidase Collagenase

DNAase

Fibrinogen → Fibrin

Coagulase

Haemolysins

Break down DNA

Kill white cells

Leucocidin

Red cell damage

Enterotoxin

Pyrogenic toxins

Diarrhoea + vomiting

TSST-1

Fever

Toxic shock syndrome

TOXIN-MEDIATED DISEASE
- Acute gastroenteritis
- Toxic shock syndrome
- Staphylococcal scalded skin syndrome

Toxin-mediated disease

SYSTEMIC INFECTION
- Bacteraemia
- Metastatic abscess
- Endocarditis
- Line-related sepsis

These non-sporing, non-motile, Gram-positive cocci forming clusters are part of the normal skin flora of humans and animals.

Classification

Staphylococci are part of the family Micrococcaceae. There are more than 26 species but only a few are associated with human disease. *Staphylococcus aureus* is the most invasive species and is differentiated from other species by possessing the enzyme coagulase.

Staphylococcus aureus

This species was once thought to be the only pathogen in the genus. Asymptomatic carriage of *S. aureus* is common and it is found in up to 40% of healthy people, in the nose, skin, axilla or perineum.

Pathogenesis

Staphylococcus aureus produces coagulase which catalyses the conversion of fibrinogen to fibrin and may help the organism to form a protective barricade. It also has receptors for the host cell surface and matrix proteins (e.g. fibronectin, collagen) that help the organism adhere. It produces extracellular lytic enzymes (e.g. lipase), which break down host tissues and aid invasion. Some strains produce potent exotoxins, which may cause a toxic shock syndrome. Enterotoxins may also be produced, causing diarrhoea.

Clinical importance

Staphylococcus aureus causes a wide range of infectious syndromes. Skin infections are favoured by warm moist conditions or when the skin is broken by disease such as eczema, by surgical wounds or by intravenous devices. Impetigo may occur in healthy skin: infection is transmitted from person to person. *S. aureus* pneumonia is rare, but may follow influenza. It progresses rapidly, with cavity formation and a high mortality. *S. aureus* endocarditis is equally rapid and destructive and may follow intravenous drug misuse or colonization of intravenous devices. *S. aureus* is the most common agent of osteomyelitis and septic arthritis (see Chapter 50).

Laboratory diagnosis

S. aureus grows readily on most laboratory media. As it is tolerant of high salt concentrations, media can be made selective in this way. Most *S. aureus* ferment mannitol: incorporation of mannitol and an indicator dye will enable them to be selected for subculture. Organisms are identified by possession of coagulase, DNAase and catalase enzymes, typical 'cluster of grapes' morphology on Gram stain, and biochemical testing. *S. aureus* can be typed by using the lytic properties of an international battery of phages or DNA restriction profiles.

Antibiotic susceptibility

The history of susceptibility of *S. aureus* is a lesson in the history of antimicrobial chemotherapy.

1 It was initially susceptible to penicillin, but β-lactamase-producing strains soon predominated.

2 Methicillin and related agents (e.g. flucloxacillin) were introduced and replaced penicillin as the drug of choice, which is still the drug of choice in sensitive strains.

3 Methicillin-resistant *S. aureus* (MRSA) emerged. Resistance is caused by possession of the *mecA* gene which codes for a low-affinity penicillin-binding protein. Some MRSA have epidemic (EMRSA) potential. Vancomycin or teicoplanin may be required for these strains.

4 Intermediate or heteroresistance to glycopeptides is emerging as an increasing issue.

5 Fully glycopeptide-resistant strains (GRSA) have been described, mediated by the *vanA vanB* genes acquired from enterococci.

Other antibiotics that are effective include linezolid, aminoglycosides, erythromycin, clindamycin, fusidic acid, chloramphenicol and tetracycline.

In methicillin-sensitive strains, first- and second-generation cephalosporins are effective. Fusidic acid may be given with another agent in bone and joint infections. Treatment should be guided by sensitivity testing.

Prevention and control

Staphylococcus aureus spreads by airborne transmission and via the hands of healthcare workers. Patients colonized or infected with MRSA or GRSA should be isolated in a side-room with wound and enteric precautions. Staff may become carriers and disseminate the organism widely in the hospital environment. Carriage may be eradicated by using topical mupirocin and chlorhexidine.

Staphylococcus epidermidis

Staphylococcus epidermidis is the most important of the coagulase-negative staphylococci (CoNS). Once dismissed as contaminants, they are now recognized as pathogens if conditions favour their multiplication.

Clinical importance

Staphylococcus epidermidis causes infection of intravenous cannulae, long-standing intravascular prosthetic devices, ventriculoperitoneal shunts and prosthetic joints. This may lead to bacteraemia or endocarditis and require the removal of the prosthesis. Biofilm production contributes to pathogenicity.

Laboratory diagnosis

S. epidermidis grows readily on laboratory media; coagulase is not produced. Speciation is by biochemical testing. DNA restriction patterns or other molecular techniques may be needed to determine whether strains are identical. *S. epidermidis* and other CoNS are common contaminants in blood cultures, requiring careful evaluation of their clinical significance.

Antibiotic susceptibility

This group of organisms is uniformly susceptible to vancomycin and usually to teicoplanin. It can be susceptible to any of the agents used for *S. aureus* infection, but this is unpredictable. Treatment must be guided by *in vitro* testing.

Staphylococcus haemolyticus

Less common than *S. epidermidis*, *S. haemolyticus* causes a similar disease pattern. It differs from *S. epidermidis* in that it causes haemolysis on blood agar. More importantly, it is naturally resistant to teicoplanin: significant infections require vancomycin therapy.

Staphylococcus saprophyticus

This coagulase-negative staphylococcus is a common cause of urinary tract infection in young women. It is distinguished by resistance to novobiocin.

14 Streptococcal infections

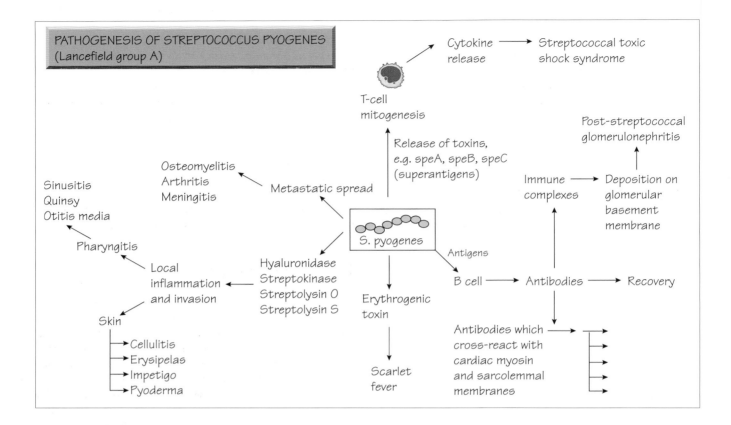

PATHOGENESIS OF STREPTOCOCCUS PYOGENES (Lancefield group A)

Cytokine release → Streptococcal toxic shock syndrome

T-cell mitogenesis

Release of toxins, e.g. speA, speB, speC (superantigens)

Osteomyelitis
Arthritis
Meningitis
Metastatic spread

Sinusitis
Quinsy
Otitis media

Post-streptococcal glomerulonephritis

Immune complexes → Deposition on glomerular basement membrane

Pharyngitis

S. pyogenes

Local inflammation and invasion ← Hyaluronidase
Streptokinase
Streptolysin O
Streptolysin S

Antigens

B cell → Antibodies → Recovery

Erythrogenic toxin

Skin
→ Cellulitis
→ Erysipelas
→ Impetigo
→ Pyoderma

Scarlet fever

Antibodies which cross-react with cardiac myosin and sarcolemmal membranes

---

These Gram-positive cocci are arranged in pairs and chains. Although facultative anaerobes, they are fastidious, requiring rich blood-containing media. Swabs from the site of infection (throat, wound, etc.) and blood culture should be taken. Colonies are distinguished by the type of haemolysis: complete (β) or incomplete (α). Biochemical and serological (Lancefield grouping) tests are used for further identification.

Streptococcus pyogenes

S. pyogenes is carried asymptomatically in the pharynx in 5–30% of the population. It is transmitted by the aerosol route and by contact. Infection is most common in children but can arise at any age.

Pathogenesis

S. pyogenes carries a group A carbohydrate antigen (Lancefield's antigen) and is surrounded by the M protein antigen, which prevents leucocyte phagocytosis. Antibodies to particular M proteins are protective against further infection with the same M type. Several toxins may be produced: for example erythrogenic toxin associated with scarlet fever, and streptococcal pyrogenic exotoxins A, B and C. The organisms attach to cells via fibronectin receptors. They can invade and survive

within cells and this may explain why pharyngeal carriage is difficult to eradicate with some antibiotics.

Clinical presentation

S. pyogenes is among the top 10 individual pathogen causes of mortality worldwide. It is associated with three types of disease.

1 Infection: it is the most common bacterial cause of pharyngitis. It also causes erysipelas, impetigo, cellulitis, wound infections and rarely, necrotizing fasciitis or pneumonia. Septicaemia may occur and result in metastatic infections (e.g. osteomyelitis). Infections are characterized by rapid onset, local tissue destruction and spreading nature. There is often significant systemic toxicity which may in part be associated with concurrent toxin production.

2 Toxin-mediated: disease in association with infection. Infection may be systemic or remain localized with systemic dissemination of exotoxins. For example, erythrogenic toxin causes scarlet fever; pyrogenic toxin-producing strains are associated with streptococcal shock and have a high mortality associated with multiple organ failure.

3 Post-infectious immune-mediated: rheumatic fever, glomerulonephritis or erythema nodosum are thought to be immune mediated because antibodies to bacterial structures cross-react with host tissues. Rheumatic fever, now uncommon in established market economies, is a major cause of long-term morbidity and mortality, particularly in areas of poverty and malnutrition.

Prevention and control

S. pyogenes can spread rapidly in surgical and obstetric wards; infected or colonized patients should be isolated in a side-room until 48 h after initiation of effective antibiotic therapy. Prompt treatment prevents secondary immune disease (e.g. rheumatic fever). Benzyl penicillin is the treatment of choice and resistance has never been reported. Amoxicillin may be used for oral therapy in less severe infections. Macrolides are an alternative for patients with allergy.

Streptococcus agalactiae

Streptococcus agalactiae (group B streptococcus) is a normal gut commensal and may be found in the female genital tract. Early (up to 1 week) perinatal infection causes pneumonia or septicaemia associated with high mortality; later infections cause meningitis. The polysaccharide antiphagocytic capsule is the main pathogenicity determinant. Prophylaxis to prevent neonatal disease is given to those mothers in labour who are febrile, known to be colonized or who have previously had an affected child. Babies of mothers with antibody to the four capsular types (Ia, Ib, II and III) are protected from infection.

Clinical features and diagnosis

Infected neonates may initially lack the classical clinical signs of sepsis, such as fever and the bulging fontanelle of meningitis. A chest X-ray may demonstrate pneumonia, and specimens of blood, CSF, amniotic fluid and gastric aspirate should be cultured. Antigen detection tests are available and can be applied to body fluids for rapid diagnosis.

Treatment and prevention

Neonatal group B streptococcal sepsis requires empirical therapy including a penicillin and aminoglycoside. Perinatal penicillin can prevent invasive infection but should be targeted at high-risk babies.

Oral streptococci
Metastatic abscesses

The '*Streptococcus milleri*' group of organisms (*S. constellatus*, *S. intermedius*, *S. anginosis*) colonize the mouth and gut. They may spread systemically, causing brain, lung or liver abscesses often as part of a mixed infection.

Infective endocarditis

Alpha-haemolytic streptococci are common causes of native valve endocarditis. Infection may be of dental origin. While good evidence is lacking, prophylaxis is recommended for at-risk patients undergoing bacteraemia-inducing dental procedures such as extraction or deep scaling.

Enterococcus spp.

Enterococci possess a group D carbohydrate cell wall antigen. Enterococci are principally commensals of the bowel. They may cause disease if they establish at other sites. Of more than 12 species, *Enterococcus faecalis* and *E. faecium* are the most common members to act as human pathogens, causing urinary tract infection, wound infection and endocarditis. Enterococci are emerging as hospital pathogens, with some species (e.g. *E. faecium*) resistant to commonly used antibiotics. Strains are usually sensitive to ampicillin/amoxycillin; however, resistance levels are increasing. Strains resistant to glycopeptides are a particular problem and may require therapy with drugs such as linezolid or pristinamycin.

15 *Streptococcus pneumoniae*, other Gram-positive cocci and the α-haemolytic streptococci

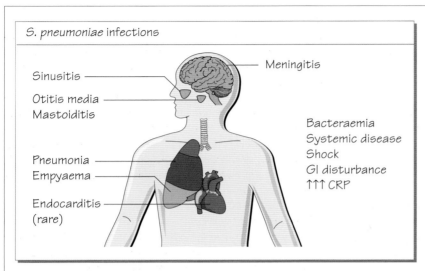

S. pneumoniae infections

Sinusitis

Otitis media
Mastoiditis

Meningitis

Bacteraemia
Systemic disease
Shock
GI disturbance
↑↑↑ CRP

Pneumonia
Empyaema

Endocarditis
(rare)

Other streptococci

Streptococcus agalactiae (group B)	Neonatal pneumonia, septicaemia, meningitis, rarely adult infection
Streptococcus equi, S. equisimilis, S. dysgalactiae, S. zooepidemicus (group C)	Endocarditis, skin infection, pharyngitis
Enterococci (group D) *E. fecalis, faecum, bovis*	UTI, endocarditis, intra-abdominal polymicrobial infection, infection in immunocompromised, ↑ resistance to glycopeptides
S. milleri group	
S. aginosis-intermedius-constellatus	Metastatic abscesses, periodontal sepsis
S. oralis, S. sanguis, S. mitis	Endocarditis
S. mutans	Dental caries

Streptococcus pneumoniae

Streptococcus pneumoniae (the pneumococcus) is a Gram-positive coccus that is typically seen in lanceolate pairs. It is sensitive to optochin and bile soluble. Haemolysis is variable. While typically partial (Alpha) strains may express several haemolysins. Pneumococcus is a major infective cause of mortality worldwide. Infection and mortality are greatest at extremes of age and in individuals with underlying disease.

Pathogenesis

Pneumococci have a polysaccharide capsule. The capsule protects pneumococci from phagocytes. There are over 90 different capsular serotypes. The capsular polysaccharide is highly antigenic and antibodies to specific types are protective. Antigenicity is type specific but there is some cross-reaction between types. Cell wall components are pro-inflammatory. The pneumococcus

also has a variety of adhesions that mediate colonization by binding to cell surface carbohydrates.

Colonization

Humans are the only host of *S. pneumoniae*; most carriage is asymptomatic. The distribution of serotypes varies with country, time and subject group. Children under 1 year of age are especially susceptible to acute pneumonia. Complement deficiency, agammaglobulinaemia, HIV infection, alcoholism and splenectomy predispose to severe infection. The capsule is the main pathogenicity determinant and there are more than 90 different types associated with varying degrees of pathogenicity and invasive potential. Toxins such as pneumolysin, neuraminidase, hyaluronidase and adhesins (e.g. pneumococcal surface protein A) are important in the pathogenesis of disease. Bacteria are able to adhere to pneumocytes and invade the bloodstream by hijacking the platelet-aggregating factor receptor pathway and through the action of pneumolysin or complement-mediated damage to the alveolus.

Clinical features

Acute otitis media, sinusitis and acute pneumonia are the most common infections. Pneumococci cause between 50 and 75% of community-acquired pneumonia of which 25–30% will be associated with bacteraemia. Direct or haematogenous spread can give rise to meningitis and, rarely, cellulitis, abscesses, peritonitis and endocarditis. Bacteraemia is an important complication with a high mortality, despite treatment (see Chapter 45). This is now the second commonest cause of meningitis in children from communities where *Haemophilus influenzae* type b (Hib) vaccination is used, and the commonest cause in adults over 40. The mortality and incidence of sequelae are high.

Antibiotic susceptibility and treatment

Once universally susceptible to penicillin, significant numbers have developed resistance through a genetically modified penicillin-binding protein gene (see Chapter 7). It is also susceptible to erythromycin, cephalosporins, tetracycline, rifampicin and chloramphenicol, but multiple drug resistance is growing. Penicillin is still the treatment of choice; cefotaxime or ceftriaxone is used for meningitis caused by less sensitive strains. Where high-level penicillin resistance occurs a glycopeptide, usually vancomycin, should be added.

Prevention and control

A polyvalent capsular polysaccharide vaccine is effective in adults but less so in immunocompromised patients and children under 2 years. A conjugate vaccine has been introduced which is immunogenic in young children.

Alpha-haemolytic streptococci
Metastatic abscesses

The 'Streptococcus milleri' group of organisms colonize the mouth and gut. They may spread systemically, causing brain, lung or liver abscesses often as part of a mixed infection. Isolation of a member of the 'S. milleri' group should prompt a thorough search for an occult abscess.

Infective endocarditis

Alpha-haemolytic streptococci such as *S. suis*, *S. oralis* and *S. mitis*, which are part of the normal flora of the mouth and gut, are common causes of native valve endocarditis. Infection may be of dental origin. While good evidence of efficacy is lacking, prophylaxis is recommended for at-risk patients (those with native valve or endocardial abnormalities and prosthetic valves), undergoing bacteraemia-inducing dental procedures such as extraction or deep scaling. *S. bovis* bacteraemia and endocarditis is associated with underlying bowel malignancy. Occasionally endocarditis is caused by nutritionally variant (pyridoxine-dependent) streptococci.

Other Gram-positive cocci

A number of other Gram-positive cocci such as *Leuconostoc* and *Pediococcus* are occasionally associated with infections, particularly in immunocompromised individuals.

Alloiococcus otitidis

Alloiococcus otitidis is a slow-growing Gram-positive coccus that produces lactic acid and has been associated with chronic otitis media with effusion in children, particularly in the chronic phase. It may be clinically more antibiotic resistant than many streptococci as it can be isolated from ear effusion despite antimicrobial therapy.

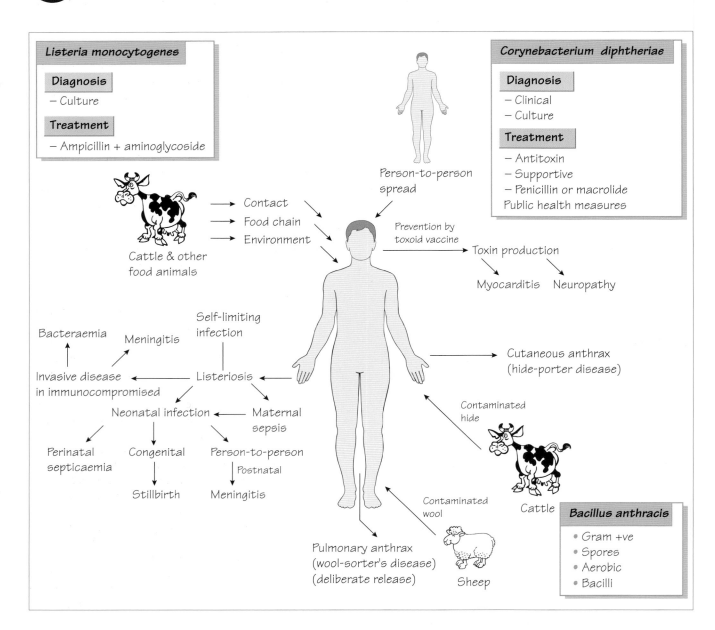

Listeria monocytogenes

Diagnosis
– Culture

Treatment
– Ampicillin + aminoglycoside

Cattle & other food animals

Contact
Food chain
Environment

Bacteraemia   Meningitis

Invasive disease in immunocompromised

Self-limiting infection

Listeriosis

Neonatal infection ← Maternal sepsis

Perinatal septicaemia   Congenital   Person-to-person
Postnatal

Stillbirth   Meningitis

Pulmonary anthrax (wool-sorter's disease) (deliberate release)

Sheep

Contaminated wool

Person-to-person spread

Prevention by toxoid vaccine

Toxin production

Myocarditis   Neuropathy

Corynebacterium diphtheriae

Diagnosis
– Clinical
– Culture

Treatment
– Antitoxin
– Supportive
– Penicillin or macrolide
Public health measures

Cutaneous anthrax (hide-porter disease)

Contaminated hide

Cattle

Bacillus anthracis
• Gram +ve
• Spores
• Aerobic
• Bacilli

Corynebacterium spp.
There are many species in this genus. All are non-sporing, non-motile and non-capsulate Gram-positive pleomorphic bacilli, arranged in an irregular pattern.

Corynebacterium diphtheriae
Corynebacterium diphtheriae is transmitted by the respiratory route or following direct contact from cutaneous lesions.

Pathogenesis
Diphtheria is caused by a strain of *C. diphtheriae* containing a bacteriophage encoding diphtheria toxin.

Clinical features and management
Infection may occur on the skin, nasopharynx or larynx. In the throat, it causes acute inflammation and necrosis shown by a green–black 'pseudomembrane' on the posterior wall of the pharynx. The membrane can cause respiratory obstruction. Diphtheria produces toxaemia: clinical severity is directly related to the degree of toxin production. The toxin lethally interrupts protein synthesis in cells, and has a direct action on the myocardium causing myocarditis, and on the peripheral nervous system causing neuropathy and paralysis. Cutaneous infection is often asymptomatic. Management is based on isolation and giving antitoxin and erythromycin. Intensive-care support may be required.

Laboratory diagnosis

Corynebacterium diphtheriae, which produces black colonies on media containing tellurite (e.g. Hoyle's), is identified by biochemical tests. Toxin production is confirmed by agar immunodiffusion (Elek's test) or detection of toxin gene by NAAT. Löffler's serum media may be useful in early isolation.

Prevention and control

Diphtheria is prevented by vaccination with a toxoid, as in the childhood immunization schedule. Immunity is long-lasting but adults may require a booster. Contacts of cases must be isolated and screened.

Corynebacterium jeikeium

Naturally resistant to most antibiotics except vancomycin, this organism causes prosthetic infection and bacteraemia in immunocompromised individuals.

Other corynebacteria

Rarely, *C. ulcerans* may cause diphtherial pharyngitis; also, *C. pseudotuberculosis* may cause suppurative granulomatous lymphadenitis. *Rhodococcus equi* has been associated with a severe cavitating pneumonia in AIDS patients.

Listeria

Gram-positive, non-sporing, motile, facultative anaerobic bacilli that can grow at low temperatures (4–10°C). *Listeria monocytogenes* is associated with human disease.

Epidemiology

Listeria spp. are found in soil or in foodstuffs where contamination by animal faeces has occurred. Cross-contamination of food products may occur. Infection follows consumption of contaminated food (e.g. soft cheeses).

Clinical features

Listeria monocytogenes may cause an asymptomatic mild, self-limiting, infectious mononucleosis-like syndrome. Alternatively, acute pyogenic meningitis, bacteraemia or encephalitis (with a high mortality) develops in patients. The mortality rate of these conditions is high, particularly in patients with reduced cell-mediated immunity. Bacteraemia in pregnancy is associated with intrauterine death, premature labour and neonatal infection similar to group B streptococci (see Chapter 43).

Laboratory diagnosis

Listeria grows readily on simple laboratory media, exhibiting a narrow zone of haemolysis on blood agar and motility at room temperature. It can be selected by incubating at low temperature but selective media allows faster isolation. Further identification is made by biochemical testing. Serological typing uses antibodies to the somatic O antigens and more recently with MLST.

Management

Listeria spp. are susceptible to ampicillin and gentamicin but resistant to the cephalosporins, penicillin and chloramphenicol. Patients with symptoms of meningitis, in whom listeriosis is a possible diagnosis, should have ampicillin incorporated into their drug regimen.

Prevention and control

Listeriosis can be prevented by good food hygiene, effective refrigeration and adequate reheating of preprepared food. Individuals who are at particular risk, such as pregnant women and the immunocompromised, should avoid high-risk foods.

Bacillus

These Gram-positive, aerobic bacilli are able to survive in adverse environmental conditions by forming spores.

Bacillus anthracis

Bacillus anthracis is a soil organism that, under certain climatic conditions, multiplies to cause anthrax in herbivores. Humans become infected from contaminated animal products. Pathogenicity depends on three bacterial antigens: the 'protective antigen' and oedema factor (both toxins); and the antiphagocytic poly D-glutamic acid capsule. Inoculation of *B. anthracis* into minor skin abrasions produces a necrotic, oedematous ulcer with regional lymphadenopathy (hide-porter's disease). Inhalation of anthrax spores develops into fulminant pneumonia and septicaemia (wool-sorter's disease). An outbreak in 2001 caused by deliberate release has led to recognition of anthrax as an agent of bioterrorism. The spores of the organisms are prepared in a way that makes them readily aerosolized so that they spread rapidly, infecting many by the respiratory route.

The diagnosis must be made in a laboratory equipped for and specialized in handling this organism. Treatment is with a penicillin, fluoroquinolone, erythromycin or tetracycline. Anthrax is prevented by animal vaccination, treatment of animal products and vaccination of humans at high risk. Antibiotic prophylaxis is used to prevent diseases associated with known exposure.

Bacillus cereus

Bacillus cereus produces a heat-stable toxin. Typically, it multiplies in par-boiled rice (during preparation of fried rice) and other contaminated food products, causing a self-limiting food poisoning. Vomiting occurs 6 h after exposure, followed by diarrhoea (18 h).

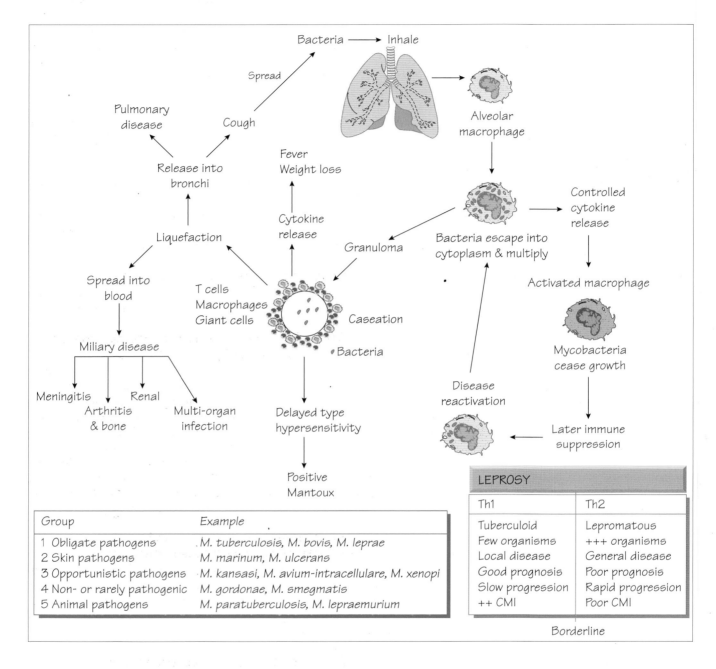

Group	Example
1 Obligate pathogens	M. tuberculosis, M. bovis, M. leprae
2 Skin pathogens	M. marinum, M. ulcerans
3 Opportunistic pathogens	M. kansasi, M. avium-intracellulare, M. xenopi
4 Non- or rarely pathogenic	M. gordonae, M. smegmatis
5 Animal pathogens	M. paratuberculosis, M. lepraemurium

LEPROSY

Th1	Th2
Tuberculoid	Lepromatous
Few organisms	+++ organisms
Local disease	General disease
Good prognosis	Poor prognosis
Slow progression	Rapid progression
++ CMI	Poor CMI

Borderline

Mycobacteria possess a lipid-rich cell wall that retains some dyes, even resisting decolorization with acid (acid-fast). There are more than 50 species: most are environmental organisms that rarely cause human infection.

Mycobacterium tuberculosis
Epidemiology and pathogenesis

Tuberculosis is spread from person to person by the aerosol route. The lung is the first site of infection. Most infections resolve with local scarring (the primary complex). Infection may disseminate from the primary focus throughout the body (miliary spread). This may resolve spontaneously or develop into localized infection (e.g. meningitis). Resistance to tuberculosis depends on T-cell function. Disease may reactivate if immunity falls (estimated 10% lifetime risk of reactivation). Immuno-compromised individuals, such as those who are HIV positive, are more likely to develop symptomatic disease.

Mycobacterium tuberculosis is ingested by macrophages but escapes from the phagolysosome to multiply in the cytoplasm. The intense immune response causes local tissue destruction (cavitation in the lung) and cytokine-mediated systemic effects (fever and weight loss). Many antigens have been identified as possible virulence determinants, including lipoarabinomannan (stimulates cytokines) and superoxide dismutase (promotes intramacrophage survival).

Clinical features

Mycobacterium tuberculosis may affect every organ of the body: it mimics both inflammatory and malignant diseases. Pulmonary tuberculosis may present with a chronic cough, haemoptysis, fever and weight loss, or as recurrent bacterial pneumonia. Untreated, the infection follows a chronic, deteriorating course. Tuberculous meningitis presents with fever and slowly deteriorating level of consciousness. Kidney infection may lead to signs of local infection, fever and weight loss, complicated by ureteric fibrosis and hydronephrosis. The lumbosacral spine is a common site of bone infection: progression may cause vertebral collapse and nerve compression. Additionally, pus may spread under the psoas sheath to appear as a groin swelling (psoas abscess). Infection of large joints may lead to a destructive arthritis. In abdominal infection, mesenteric lymphadenopathy and chronic peritonitis may present as fever, weight loss, ascites and intestinal malabsorption. Disseminated infection (miliary disease) can occur without evidence of active lung infection.

Laboratory diagnosis

Specimens are stained by Ziehl–Neelsen's method, then cultured on lipid-rich (egg-containing) medium with malachite green (Löwenstein–Jensen, L–J) to suppress other organisms. Growth can be detected more quickly in semi-automated continuous monitoring systems. Susceptibility is tested on slopes of L–J medium or in automated systems. Nucleic acid amplification testing (NAAT) aids rapid diagnosis. NAAT and sequence determination of the *rpoB* gene permits rapid identification of MTb complex and diagnosis of rifampicin resistance. *Mycobacterium tuberculosis* can be typed by a restriction fragment length polymorphism (RFLP) method. New measurements of cytokine responses in peripheral blood mononuclear cells are being evaluated to assess their ability to differentiate exposure from active disease.

Treatment and prevention

The standard regimen for pulmonary infection is rifampicin and isoniazid for 6 months, with ethambutol and pyrazinamide for the first 2 months. Regimes for other sites are similar, taking into account drug penetration (e.g. into CSF). There is a rising trend of multidrug-resistant tuberculosis (MDRTB). A history of previous incomplete treatment, residence in a country with high incidence of MDRTB or failure to respond clinically to an adequate regimen suggest the possibility of MDRTB. Treatment for MDRTB is with a combination of second-line agents such as aminoglycosides, fluoroquinolones, ethionamide or cycloserine, and is guided by susceptibility tests.

Vaccination with attenuated bacille Calmette–Guérin (BCG) strain may protect against miliary spread, but trials in some countries have shown no benefit. Patients at high risk of developing tuberculosis may be given prophylaxis with isoniazid and rifampicin. HIV patients may benefit from long-term prophylaxis with rifabutin or clarithromycin.

Mycobacterium leprae

Mycobacterium leprae cannot be cultivated in artificial medium. The organism attacks peripheral nerves, causing anaesthesia. Digital destruction and deformity follow, leaving the patient severely disabled. The end result of infection depends on the individual immune response, forming a spectrum from 'tuberculoid' dominated by a Th1 response, through 'borderline', to 'lepromatous' dominated by a Th2 response. Patients with tuberculoid disease have a strong cell mediated immune response, many granulomas and a paucity of bacteria in tissues associated with trophic nerve damage, whereas patients with lepromatous disease have poor cell-mediated immunity (CMI), no granuloma, and generalized disease (leonine facies, depigmentation and anaesthesia).

Diagnosis is by Ziehl–Neelsen's stain of a split-skin smear and histological examination of skin biopsy. Treatment with rifampicin, dapsone and clofazimine renders the patient noninfectious rapidly, but cannot alter nerve damage and deformity, which must be managed by remedial surgery.

Non-tuberculous mycobacteria

Different species may cause localized or disseminated disease in immunocompromised patients. Some may infect grafts.

Mycobacterium avium–intracellulare complex (MAIC)

This complex includes *Mycobacterium avium*, *M. intracellulare* and *M. scrofulaceum*. Some are natural pathogens of birds, others are environmental saprophytes. A common cause of mycobacterial lymphadenitis in children, they also cause osteomyelitis in immunocompromised patients and chronic pulmonary infection in the elderly. In advanced HIV disease, they cause disseminated infection and bacteraemia. MAIC are naturally resistant to many antituberculosis agents; multidrug regimens should be used, including rifabutin, clarithromycin and ethambutol. Lymphadenitis may require surgery.

Mycobacterium kansasi, M. malmoense and M. xenopi

These species cause an indolent pulmonary infection resembling tuberculosis in individuals predisposed by chronic lung disease such as bronchiectasis, silicosis and obstructive airways disease. Initial therapy with standard drugs may have to be adjusted following susceptibility tests.

Mycobacterium marinum and M. ulcerans

Mycobacterium marinum causes a chronic granulomatous infection of the skin and is acquired from rivers, poorly maintained swimming pools or fish tanks. It is characterized by encrusted pustular lesions. *M. ulcerans* infection is associated with farming in Africa and Australia. The lower limb is usually affected with a papular lesion, which ulcerates and may destroy underlying tissue including bone.

18 *Clostridium*

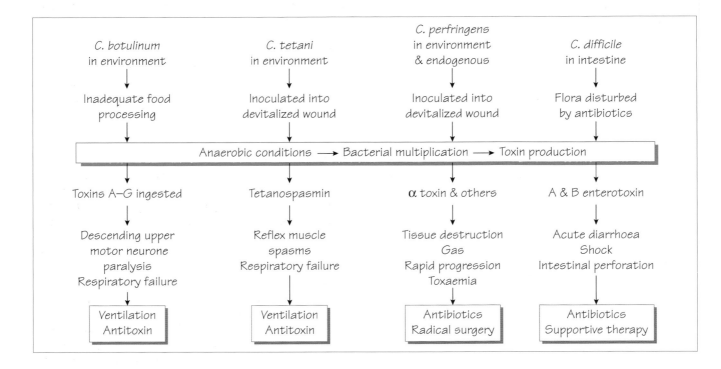

Clostridium spp. are obligate, anaerobic, spore-forming, Gram-positive bacilli. Of the 80 species, few are human pathogens. Their normal habitat is in soil, water and the intestinal tract of humans and animals. They cause disease as a consequence of toxin production.

Pseudomembranous colitis

This is caused by *Clostridium difficile* and is an important disease in modern hospitals.

Epidemiology and pathogenesis

Clostridium difficile is found in the human intestine, especially in hospitalized patients whose gut microflora has been disturbed by antibiotics. It produces enterotoxins A and B, causing fluid secretion and tissue damage. Neonates commonly carry the organism and toxin without ill-effect; susceptibility increases with age. Some strains (e.g. type O27) with enhanced toxin production are associated with more severe disease.

Clinical features

Typically, the patient passes more than three loose or unformed stools per day, with a history of previous antibiotic exposure. Abdominal pain may develop and sigmoidoscopy will reveal pseudomembranes: small white–yellow plaques situated on the mucosal surface of the rectum and sigmoid colon. The diagnosis is confirmed by the laboratory demonstration of toxin in the stool by EIA or tissue culture. Severe forms are accompanied by toxic megacolon, bowel perforation and systemic toxicity associated with a high mortality.

Treatment and prevention

The inciting agent should be stopped and the patient given oral metronidazole for 10 days. Oral vancomycin is an alternative. Relapse occurs in up to 25% of patients. Patients with pseudomembranous colitis should be isolated from other patients, using enteric precautions. In severe disease combination therapy, immunoglobulin and surgery should be considered. Infection control measures are important.

Tetanus caused by *Clostridium tetani*
Epidemiology and pathogenesis

Infection occurs in wounds deep enough to produce anaerobic conditions. *C. tetani* produces tetanospasmin which prevents release of the inhibitory transmitter γ-aminobutyric acid (GABA), resulting in muscle spasms. Neonatal tetanus, which may occur if the umbilical stump is contaminated after delivery, is an important cause of death in developing countries. Tetanus is rare in developed countries (0.2 cases per million), usually occurring in older patients in whom immunity has declined. There is often a history of a trivial gardening injury.

Clinical features

Spastic paralysis and muscle spasms may develop at the site of the lesion and if untreated become generalized. Perioral muscle spasm leads to the risus sardonicus, and spasm of the spinal muscles and legs lead to opisthotonus (when the head and heels are bent back towards each other). Spasms are painful, and may be stimulated by light or sudden noise. There may be respiratory embarrassment and secondary bacterial pneumonia. Diagnosis is based on history and clinical features; isolation of the organism is not diagnostic.

Treatment and prevention

Treatment is with muscle relaxants and limiting further toxin activity by the use of human tetanus immunoglobulin and antibiotics. Ventilation and treatment of secondary pneumonia may be required.

Infants are protected by passive immunity from their mothers, and develop active immunity when they receive tetanus toxoid as part of their childhood immunization course. Boosters are given at school entry and every 10–15 years. Unvaccinated patients with tetanus-prone wounds should receive antibiotics and human tetanus immunoglobulin, followed by a course of vaccination.

Botulism

Clostridium botulinum has seven types, named A–G, based on biochemical testing and toxin serotype. Serotypes A, B and E are most commonly implicated in human disease.

Epidemiology and pathogenesis

Clostridium botulinum can contaminate foods such as meat or vegetables. Incomplete heat treatment in the canning or bottling process allows this organism to survive and produce toxin. Botulinum toxin, a neurotoxin, inhibits the release of neuro-transmitters. Clinically, there are three forms of the disease: food intoxication, wound botulism and infant botulism. Wound and infant botulism may lead to systemic toxaemia.

Clinical features

A descending flaccid paralysis, beginning with the cranial nerves, develops within 6 h of ingestion of toxin-contaminated food. Patients develop dysphagia and blurred vision, followed by more general paralysis, but are not confused and sensory function is normal. Infants appear floppy and listless, are con-stipated and have generalized muscle weakness. Diagnosis is based on clinical features and history of ingestion of suspect food. Toxin may be demonstrated in faeces and serum by EIA.

Treatment and prevention

Treatment is with specific antitoxin and ventilatory support. Penicillin is also used to eradicate the organism. The disease is prevented by adequate process control in the food-processing industry and home preservation.

Gas gangrene

Clostridium perfringens is the organism most commonly associated with gas gangrene, but *C. septicum*, *C. novyi*, *C. his-tolyticum* and *C. sordellii* can also be implicated. *Clostridium perfringens* is capsulate and produces a range of toxins, of which lecithinase C (α-toxin) is the most important.

Epidemiology and pathogenesis

Typically, gas gangrene develops when a devitalized wound becomes contaminated with spores from the environment. The spores germinate and organisms multiply in the ischaemic conditions, releasing toxins which cause further tissue damage. Progression is rapid. Mixed infections occur at injection sites in injecting drug users.

Clinical features

Gangrene develops within 3 days of injury. The wound is painful; the skin becomes tense with an underlying blue dis-coloration, foul smell and crepitus. Toxaemia will produce circulatory shock. The diagnosis is made clinically. Microscopy of stained smears may reveal necrotic material, a few inflam-matory cells and large Gram-positive bacteria.

Treatment and prevention

Treatment depends on debridement of devitilized tissue and intravenous antibiotics. Hyperbaric oxygen may also be beneficial. The condition may be prevented by good manage-ment of potentially infected devitalized wounds.

Clostridium perfringens food poisoning

This condition is typically associated with meat meals which cool slowly and are reheated. Surviving clostridia release toxin in the stomach when they form spores, leading to nausea, vomiting and diarrhoea. An EIA to detect toxin in faeces is available. Rarely, gut infection with clostridia causes severe enteritis.

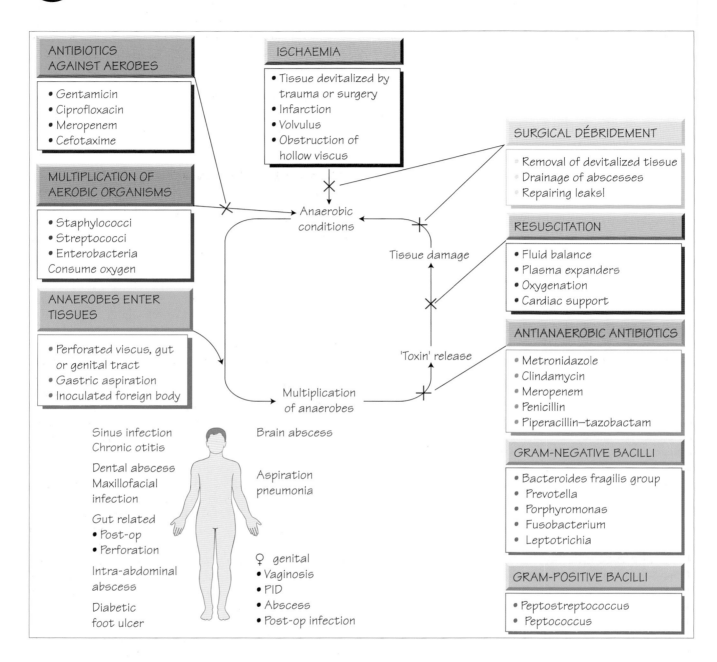

Non-sporing anaerobes

Non-sporing anaerobes form the major part of the normal human bacterial flora, outnumbering all other organisms in the gut by a factor of 10^3. They are also found in the genital tract, oropharynx and skin.

Anaerobic sepsis
Pathogenesis

Infection with non-sporing anaerobes is usually endogenous. Organisms from the normal flora escape into a sterile site (e.g. following perforation of the large intestine). Conditions that allow anaerobic growth may arise after ischaemia (e.g. a strangulated hernia), or when facultative bacteria produce anaerobic conditions by the action of their metabolism (e.g. in deep skin ulcers or intraperitoneal infection).

Once established, anaerobic multiplication is promoted by release of toxic metabolic products and proteolytic enzymes. The toxic products of inflammatory cells may also be released, such as reactive oxygen intermediates. These exacerbate tissue damage, allowing further anaerobic invasion and multiplication. The cycle of anaerobic sepsis, if not halted, rapidly leads to septicaemia and death.

Clinical importance

Generalized intra-abdominal sepsis may follow spontaneous bowel perforation or elective surgery, leading to abscess formation (e.g. abdominal or liver abscesses).

Non-sporing anaerobes play a prominent part in sepsis of the female genital tract. These infections are often secondary to septic abortion, prolonged rupture of the membranes, complicated caesarean section or retained products of conception. They are directly implicated in pelvic inflammatory disease (PID), while imbalance in the anaerobic flora of the vagina may lead to the non-specific vaginosis syndrome (see Chapter 49).

Anaerobes are often part of polymicrobial liver abscesses and biliary sepsis.

Pneumonia following aspiration or associated with carcinoma or foreign-body obstruction has a significant anaerobic component. Such infections can develop into a lung abscess.

Brain abscesses often have an important anaerobic component, as does chronic paranasal suppuration, such as in chronic otitis media and chronic sinusitis.

Anaerobes may colonize chronic skin ulcers, such as lower leg ulceration in elderly people. The less common tropical ulcer is caused by *Fusobacterium ulcerans*.

Laboratory diagnosis

These organisms are nutritionally fastidious and oxygen is toxic to them. Specimens should be plated directly in theatre or at the bedside, or transported to the laboratory rapidly in an anaerobic transport system. Pus rather than swabs (which dry out quickly) should be sent. Specimens are inoculated into a fluid enrichment medium (e.g. Robertson's cooked meat, fastidious anaerobe medium) and onto blood-containing media, some of which contain antibiotics to inhibit growth of aerobes. Plates must be incubated under strict anaerobic conditions.

Anaerobic species are identified on the basis of their Gram reaction, their growth on bile- or dye-containing medium or their biochemical reactions, or by studying the end-products of metabolism using gas–liquid chromatography.

Antibiotic susceptibility

Almost all anaerobes are susceptible to metronidazole, although resistance has been reported. Other active agents include meropenem, piperacillin–tazobactam, clindamycin, chloramphenicol, penicillin and erythromycin. (NB. Penicillin and erythromycin are not active against *Bacteroides fragilis*, the anaerobe most commonly isolated from abdominal sepsis.)

Management

Effective management of anaerobic sepsis depends on a dual approach—surgery and antimicrobial agents. Surgical procedures may include closing perforations, resecting a gangrenous hernia, debriding non-viable tissue from ulcers, draining abscesses and treating coexisting infection. Metronidazole is the most commonly used anti-anaerobic agent.

Prevention and control

The risk of anaerobic infection in elective surgery can be reduced by good operative technique and perioperative antibiotics with anti-anaerobic activity.

Pathogens of anaerobic sepsis

Bacteroides fragilis is the most common agent of serious anaerobic sepsis. It is penicillin resistant by virtue of its β-lactamase production. It also produces a protease, DNAase, heparinase and neuraminidase. It has an antiphagocytic capsule and inhibits the phagocytosis of facultative organisms, promoting the development of synergistic infections. *Bacteroides fragilis* is typically associated with postoperative sepsis in abdominal and gynaecological surgery. It also contributes to the polymicrobial flora found in cerebral, hepatic and lung abscesses.

Prevotella melaninogenicus and fusobacteria are found chiefly in the oral cavity. They are associated with periodontal disease, gingivitis, dental abscess, sinus infection, cerebral and lung abscesses and necrotizing pneumonia. They are also found in association with *Borrelia vincentii* in Vincent's angina, and in ulcerative diseases such as cancrum oris (Ludwig's angina), both of which affect the head and neck. They may contribute to anaerobic cellulitis.

Peptococcus and *Peptostreptococcus* are the only anaerobic Gram-positive cocci that are regularly found in human specimens. They are usually found in mixed infections, such as dental sepsis, cerebral or lung abscesses, and soft-tissue and wound infections. They are also associated with necrotizing fasciitis, where a mixed infection of anaerobic cocci, facultative streptococci and possibly also *Staphylococcus aureus* progresses rapidly, with destruction of the skin and deeper tissues, leading to septicaemia and death.

Actinomyces-like bacteria rarely are associated with chronic abscesses following dental sepsis, lung abscesses, gut perforation and infection of intrauterine devices. Long-term penicillin is usually effective.

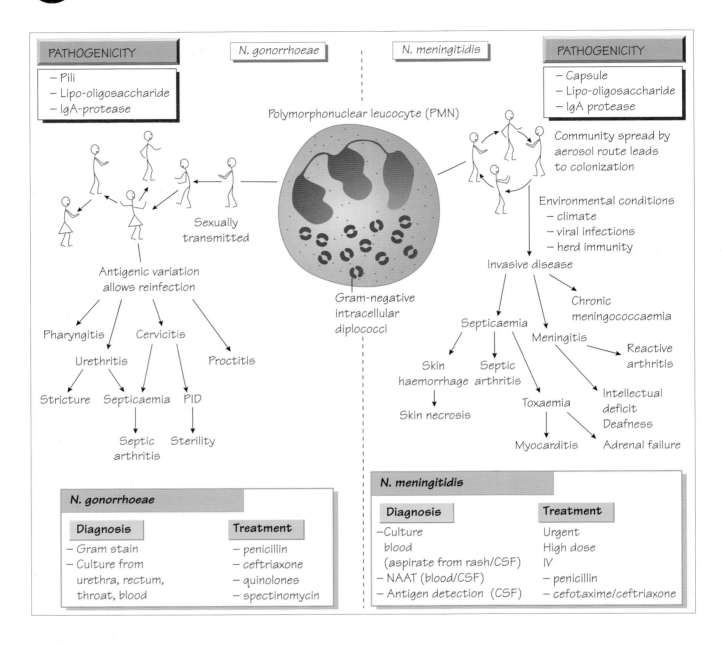

PATHOGENICITY

– Pili
– Lipo-oligosaccharide
– IgA-protease

N. gonorrhoeae

N. meningitidis

PATHOGENICITY

– Capsule
– Lipo-oligosaccharide
– IgA protease

Polymorphonuclear leucocyte (PMN)

Community spread by aerosol route leads to colonization

Sexually transmitted

Environmental conditions
– climate
– viral infections
– herd immunity

Antigenic variation allows reinfection

Gram-negative intracellular diplococci

Invasive disease

Pharyngitis   Cervicitis

Urethritis   Proctitis

Stricture   Septicaemia   PID

Septic arthritis   Sterility

Septicaemia

Skin haemorrhage   Septic arthritis

Skin necrosis

Chronic meningococcaemia

Meningitis

Reactive arthritis

Toxaemia

Intellectual deficit Deafness

Myocarditis   Adrenal failure

N. gonorrhoeae

Diagnosis	Treatment
– Gram stain	– penicillin
– Culture from urethra, rectum, throat, blood	– ceftriaxone – quinolones – spectinomycin

N. meningitidis

Diagnosis	Treatment
–Culture blood (aspirate from rash/CSF)	Urgent High dose IV
– NAAT (blood/CSF)	– penicillin
– Antigen detection (CSF)	– cefotaxime/ceftriaxone

Neisseria gonorrhoeae

Infection with *N. gonorrhoeae* is most common between 15 and 35 years of age. It is almost exclusively spread by sexual contact. Antigenic variation of the gonococcal pili means that recovery from infection provides no immunity, so reinfection is possible.

Pathogenesis

Neisseria gonorrhoeae is a Gram-negative coccus that adheres to the genitourinary epithelium via pili. It invades the epithelial layer and provokes a local acute inflammatory response.

Clinical features

Neisseria gonorrhoeae causes acute painful urethritis and urethral discharge. Female infection (cervicitis) is often asymptomatic or may be associated with vaginal discharge. Pelvic inflammatory disease may develop. Pharyngeal infection causes pharyngitis. Rectal infection or proctitis presents with a purulent discharge. Infection can be complicated by bacteraemia, septic or reactive arthritis of the large joints, or pustular skin rashes. Late complications include female infertility and male urethral stricture.

Diagnosis

Pus from urethra, cervix, rectum or throat should be plated directly, or transported rapidly to the laboratory in specialized transport medium. Identification is based on biochemical and serological methods. Expression of β-lactamase can be detected by a rapid colorimetric test. NAAT methods are more sensitive than culture.

Treatment and prevention

Treatment must be given before susceptibility results are available. This is usually penicillin: alternatives are ciprofloxacin, ceftriaxone or spectinomycin. Emergence of resistance is a problem so local sensitivity data are important. Gonorrhoea can be prevented by avoiding high-risk sexual contacts and using barrier contraception. Contacts of infected individuals should be traced and treated. Antigenic variation of the pili precludes vaccine development at present.

Neisseria meningitidis
Epidemiology

Carriage of *N. meningitidis* is common; actual disease only develops in a few individuals. Infection is most common in the winter. Epidemics occur every 10–12 years. In Africa, severe epidemics occur in the 'meningitis belt' where the incidence can rise to 1000 cases per 100 000 each year. Most invasive infections are caused by serogroups A, B or C. Group B infection is now the commonest. The incidence of group C infection has reduced in communities where vaccination is now routine.

Pathogenesis and clinical features

The main pathogenicity determinant of *N. meningitidis* is the antiphagocytic polysaccharide capsule. Meningococci cross mucosal epithelium by endocytosis and the capsule allows survival in the bloodstream. Lipo-oligosaccharide activates complement activation and cytokine release, leading to shock and disseminated intravascular coagulation (DIC).

Meningococcal meningitis is characterized by fever, neck stiffness and reduced consciousness. The petechial rash, a sign of septicaemia, may be present without other signs of meningitis. Septic or reactive arthritis may develop.

Diagnosis and treatment

The diagnosis is usually made clinically and confirmed by culture of blood, aspirate from the rash and CSF. Rapid antigen detection or NAAT on CSF and blood are sensitive and reliable.

Infection is life-threatening and rapidly progressive: treatment should not await laboratory confirmation or hospitalization. Intravenous penicillin G (intramuscular in the community setting) is the antibiotic of choice, but there have been reports of meningococci with reduced susceptibility in other countries. Cefotaxime is an alternative. Treatment does not eradicate carriage so the patient should be given 'prophylaxis' following recovery.

Prevention

A polysaccharide vaccine is available against serogroups A and W135. A serogroup C protein conjugate vaccine is more than 90% efficient. An effective vaccine against serogroup B is not available yet. Close contacts of patients with meningococcal meningitis should be given 'prophylaxis' with rifampicin or ciprofloxacin.

Moraxella catarrhalis

This Gram-negative coccobacillus is usually a commensal of the upper respiratory tract. It is associated with otitis media, sinusitis and lower respiratory tract infection in children or patients with chronic pulmonary disease. It usually produces β-lactamase.

Haemophilus

Haemophilus spp. are small Gram-negative coccobacilli that are dependent for growth on blood factors X and/or V. They colonize mucosal surfaces. *Haemophilus influenzae* and *H. ducreyi* are the main pathogenic species.

Haemophilus influenzae

Haemophilus influenzae expresses an antiphagocytic polysaccharide capsule of which there are six types (a–f). It also expresses a lipopolysaccharide and an IgA$_1$ protease. Septicaemia, meningitis and osteomyelitis are usually associated with type b in individuals who have not been vaccinated.

Clinical features

Infection occurs in preschool children, causing pyogenic meningitis, acute epiglottitis, septicaemia, facial cellulitis or osteomyelitis. Non-capsulate strains are usually commensal in the nasopharynx, but may cause adult otitis media, sinusitis, and chest infection in patients with obstructive airways disease.

Laboratory diagnosis

Antigen detection provides rapid diagnosis in meningitis. Culture of CSF, sputum, blood or pus must be on 'chocolate' agar incubated in 5% carbon dioxide. *Haemophilus influenzae* is identified by X and V dependence.

Treatment and prevention

Many *H. influenzae* express a β-lactamase and are ampicillin resistant. Co-amoxiclav, clarithromycin, tetracycline or trimethoprim can be used. Severe infections are treated with β-lactam-stable cephalosporin.

A protein-conjugated polysaccharide vaccine against type b has almost eradicated childhood infection. Non-capsulate *Haemophilus* is ubiquitous: predisposed patients cannot avoid infection.

Haemophilus ducreyi

Haemophilus ducreyi is transmitted sexually and causes painful, irregular, soft genital ulcers (chancroid). There is associated lymphadenopathy, and suppurating inguinal lymph nodes may lead to sinus formation. Infection is more common in developing countries and facilitates the transmission of HIV.

Transmission is controlled by treatment with erythromycin or co-amoxiclav, and contact tracing.

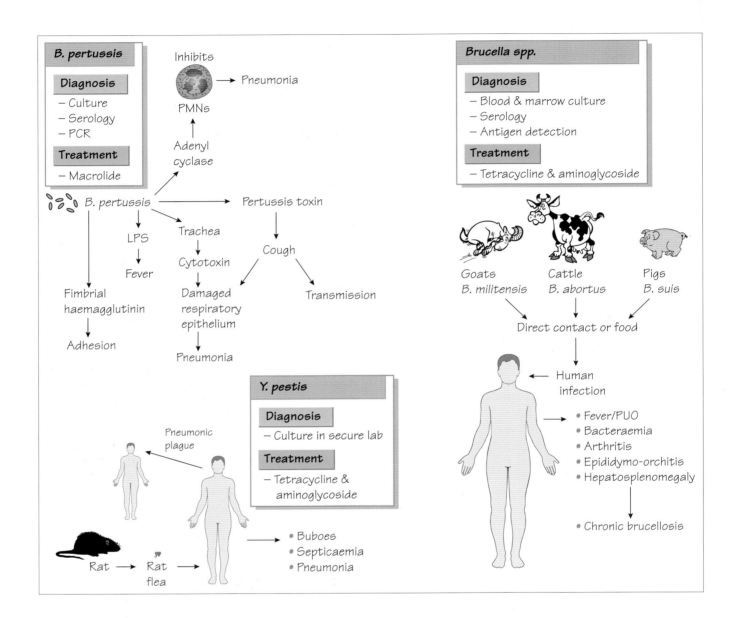

Bordetella spp.

Bordetella pertussis and *B. parapertussis* can cause whooping cough. In the absence of an adequate vaccination campaign, epidemics of whooping cough occur in children every 4 years. Asymptomatic or unrecognized infection in adolescents and young adults maintains the cycle of infection in the human population.

Pathogenesis

Bordetella pertussis expresses fimbriae that aid adhesion, and a number of exotoxins including pertussis toxin, adenyl cyclase and tracheal cytotoxin. There is a complex interaction with the cells of the respiratory tract which results in thickened bronchial secretions and paroxysmal cough. Secondary respiratory tract infection, apnoea induced by prolonged paroxysms of coughing, and raised intracranial pressure are the main complications.

Clinical features

A 2-week catarrhal phase precedes the characteristic whooping cough: repeated, prolonged bouts of coughing followed by a deep inspiratory whoop. The whoop is often absent in very young children and adults. It is frequently associated with vomiting and subconjunctival haemorrhage. The coughing phase can last for up to 3 months. Young children feed poorly and lose weight.

As spasms are more frequent at night, loss of sleep for parents and child has a significant contribution to the morbidity of the disease. Infection is often complicated by secondary pneumonia and otitis media.

Laboratory diagnosis

Specimens are obtained using a pernasal swab. Inoculated onto charcoal blood agar and incubated for up to 5 days, they develop into small pearly colonies. Positive cultures are more likely in the catarrhal and early coughing phase. Diagnosis can also be made by antigen detection, EIA or NAAT.

Treatment

Erythromycin is thought to decrease infectivity and shorten symptoms if given early during the catarrhal phase. Symptomatic support and early treatment of secondary infections are the mainstay of treatment.

Prevention and control

A whole-cell, killed vaccine is effective when high vaccine coverage of the community is obtained. Long-standing concerns about the safety of this vaccine remain unproved. Subcellular vaccines are used in some countries.

Brucella spp.

Brucella melitensis, *B. abortus* and *B. suis* have goats, cattle and pigs, respectively, as their main hosts. They are aerobic or capnophilic, requiring serum-containing medium to grow.

Brucella infection spreads to humans through direct contact with domesticated animals or their products (e.g. unpasteurized milk). Veterinarians, farmers and abattoir workers are at increased risk of infection.

Pathogenesis

Brucellae are able to survive inside cells of the reticuloendothelial system using a superoxide dismutase and nucleotide-like substances to inhibit the intracellular killing mechanisms of the host.

Clinical features

Intermittent high fever is characteristic of the early stages of infection, giving rise to the old name 'undulant fever'. It is associated with myalgia, arthralgia and lumbosacral tenderness. Acute infection can be complicated by septic arthritis, osteomyelitis and epididymo-orchitis. Without treatment a chronic infection develops, which may resolve or continue to give symptoms, often accompanied by psychiatric complaints, for many years.

Laboratory diagnosis

Culture of blood and bone marrow is diagnostic. Incubation, in a high containment facility, must be extended for up to 3 weeks. In chronic disease, culture is less likely to be positive. Bacterial agglutination is used as a simple screening test: positive results may be confirmed by EIA to detect both IgG and IgM.

Treatment and prevention

Optimal treatment is with tetracycline for 1 month. Streptomycin should be added for patients with complications. Transmission by food can be prevented by pasteurization. Appropriate animal husbandry techniques can reduce the occupational risk of infection. An animal vaccine is available but is not sufficiently safe for human use. Animal control measures have eradicated brucellosis from farms in many countries.

Francisella tularensis

This pathogen of rodents and deer can be found in North America and northern Europe. Infection is spread by the aerosol route, by direct contact with feral animals, or by tick bite. This rare infection occurs mainly among campers and hunters. Infection may be ocular or localized to the skin, with regional lymphadenopathy. Systemic infection gives a syndrome that resembles typhoid, with 5–10% mortality. Diagnosis is by serology or by culture. Treatment is with tetracycline.

Yersinia
Yersinia pestis

Yersinia pestis infection is described in Chapter 52.

Yersinia enterocolitica

This organism, morphologically and biochemically similar to *Y. pestis*, causes acute enteritis, mesenteric adenitis and, rarely, septicaemia. It is transmitted to humans in food and water. Infection may be complicated by polyarthritis and erythema nodosum. Patients with iron overload syndromes are especially susceptible. Diagnosis is made by isolation from faeces, blood or lymph node, or detection of antibodies. Treatment with ciprofloxacin or co-trimoxazole is indicated in serious infection; tetracycline is an alternative therapy.

Yersinia pseudotuberculosis

This organism can cause mesenteric adenitis that mimics appendicitis.

Bartonella spp.

Bartonella are small Gram-negative bacteria which can invade host red, epithelial and bone-marrow cells. *Bartonella henselae* is responsible for cat-scratch disease (Chapter 52) and bacillary angiomatosis, a febrile illness associated with a red papular rash commonly found in HIV-infected patients. It is also associated with endocarditis. *Bartonella quintana* is responsible for trench fever, a relapsing febrile illness now found in the homeless. *Bartonella bacilliformis* infection can cause Oroya fever, an acute febrile haemolytic anaemia or mild fever with body pain, nausea and headache. Diagnosis is by culture, but NAAT and sequencing is more sensitive.

22 Pathogenicity of enteric Gram-negative bacteria

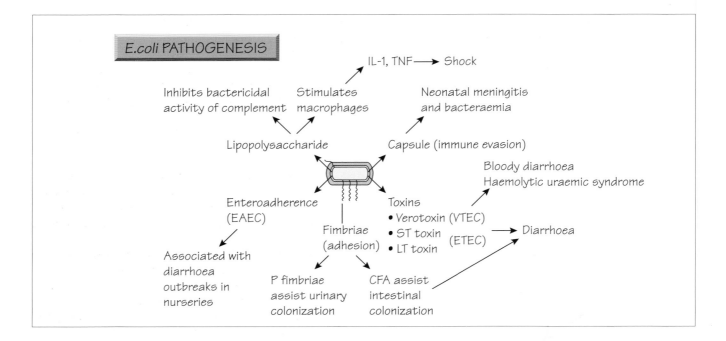

The Enterobacteriaceae are a large family (more than 20 genera and 100 species) of facultatively anaerobic, Gram-negative bacilli that are easily cultured, reduce nitrate and ferment glucose. The wide diversity of named species is in part because they are easy to grow and study in the laboratory, and also because they are capable of causing a wide variety of clinical syndromes.

Habitat and transmission

The Enterobacteriaceae are almost ubiquitous organisms. They are found as a major component of the normal flora of animals and humans, and can be found in the environment in water and soil. Transmission is both from other animals or humans and from the inanimate environment. Many infections arise from the body's normal flora when opportunities are provided by medical, surgical or other therapy. The proportions of different species vary with body site, with shifts in states of health and disease, and in response to antimicrobial pressure.

Pathogenicity
Capsules

Many species produce extracellular capsular polysaccharides (e.g. *Klebsiella* spp., *Escherichia coli* and *Salmonella typhi*). *Salmonella typhi* possesses a capsule or Vi (virulence) antigen; therefore vaccine containing the Vi antigen is protective against typhoid. *Escherichia coli* K1 is the most common type of *E. coli* isolated from neonatal meningitis and septicaemia. There are biochemical and structural similarities between *E. coli* K1 and *N. meningitidis* group B and human central nervous system antigens which may give these pathogens an advantage.

Lipopolysaccharide

The lipopolysaccharide (LPS) molecule consists of a central lipid A and oligosaccharide core, and a long straight or branched polysaccharide 'O' antigen. It is located in the bacterial outer membrane and is responsible for resistance to the bactericidal activity of complement. The lipid A core stimulates host macrophages to produce cytokines, such as interleukin-1 and tumour necrosis factor (TNF), responsible for the fever, shock and metabolic acidosis associated with severe sepsis. Some clinical syndromes are associated with particular O antigens; for example *E. coli* O157 may produce verotoxin causing haemolytic–uraemic syndrome, other O types are associated with urinary tract infection or diarrhoea. However, these are merely temporal relationships between a variety of bacterial characteristics that include an O antigen and a particular virulence determinant.

Urease

Proteus spp. express a potent urease that splits urea. In the urinary tract, urea lowers the pH, which in turn allows calcium and phosphates to precipitate and is thus associated with the formation of renal stones (see Chapter 49).

Fimbriae

Fimbriae or pili are bacterial organelles that allow attachment to host cells and are important in promoting colonization in environments where the organisms could be dislodged, such as the ureter. *Escherichia coli* expressing mannose-binding fimbriae are associated with lower urinary tract infections and cystitis, whereas those that express P fimbriae are associated with pyelonephritis and septicaemia. In the intestine, *E. coli* that express different fimbriae (colonization factor antigens, CFAs) have been associated with diarrhoea.

Toxins

Enterotoxigenic *E. coli* (ETEC)
Enterotoxigenic *E. coli* produce LT toxin and ST toxin. These toxins act on the enterocyte to stimulate fluid secretion, resulting in diarrhoea. LT toxin shares 70% homology with cholera toxin. It is heat labile (LT), and like cholera toxin increases local cyclic adenosine monophosphate (cAMP) in the enteric cell. ST is heat stable and stimulates cyclic guanyl monophosphate. *Escherichia coli* possessing these enterotoxins are associated with travellers' diarrhoea: a short-lived, watery diarrhoeal disease.

Enteroaggregative *E. coli* (EAggEC)
Some strains of *E. coli* are able to attach to, and cause aggregation of, enteric cells. They do not invade the cells, and are known as enteroaggregative *E. coli* (EAggEC) and are able to cause chronic diarrhoea. They are covered with fibrillar structures which are presumed to mediate adherence. Strains express an ST-like toxin or a haemolysin-like toxin.

Enteropathogenic *E. coli* (EPEC)
The *E. coli* with this characteristic were the first *E. coli* recognized as primary pathogens causing outbreaks of diarrhoea in preschool nurseries. Adherence is associated with loss of microvilli and is caused by rearrangement of host cell actin.

Enterohaemorrhagic *E. coli* (EHEC)
These strains produce a verotoxin named because of its activity on vero cells *in vitro*. The haemorrhagic diarrhoea that they cause can be complicated by haemolysis and acute renal failure: the haemolytic–uraemic syndrome. This organism is commensal in cattle and is transmitted to humans through hygiene failure in abattoirs and food production. A similar toxin (Shiga toxin) is a major virulence determinant in *Shigella dysenteriae*.

Genetic exchange
The Enterobacteriaceae can gain DNA rapidly from other organisms through transposons, integrons or plasmids. This enables antibiotic resistance genes to spread from one species to another. In the hospital environment the survival of antibiotic-resistant strains is favoured. In some hospitals there have been outbreaks of multidrug-resistant *Klebsiella pneumoniae* in intensive-care units. The Enterobacteriaceae have also been able to gain pathogenicity determinants by genetic exchange. Acquisition of a series of connected genes can occur and these are known as pathogenicity islands. *Salmonella* have gained a series of genes in this way, enabling them to invade intestinal cells.

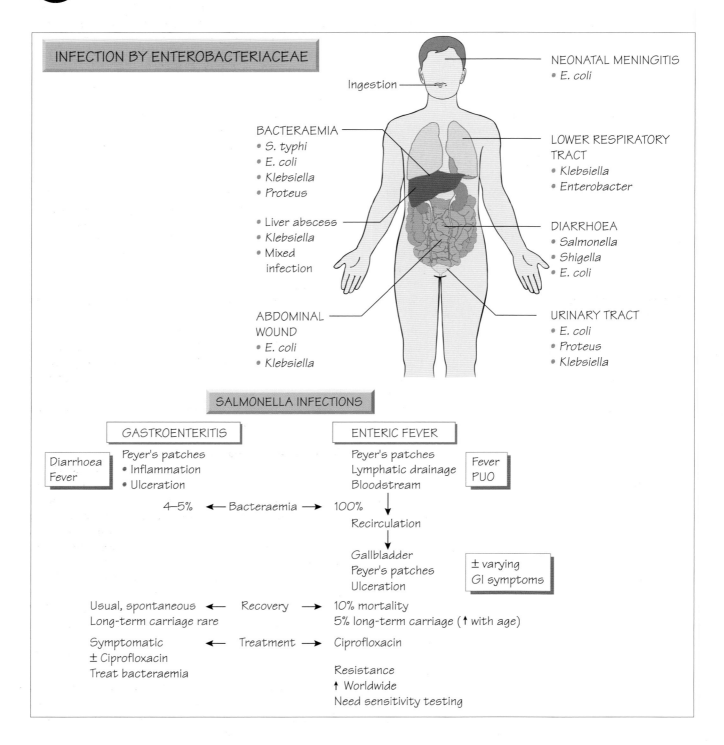

INFECTION BY ENTEROBACTERIACEAE

Ingestion

NEONATAL MENINGITIS
• E. coli

BACTERAEMIA
• S. typhi
• E. coli
• Klebsiella
• Proteus

LOWER RESPIRATORY TRACT
• Klebsiella
• Enterobacter

• Liver abscess
• Klebsiella
• Mixed infection

DIARRHOEA
• Salmonella
• Shigella
• E. coli

ABDOMINAL WOUND
• E. coli
• Klebsiella

URINARY TRACT
• E. coli
• Proteus
• Klebsiella

SALMONELLA INFECTIONS

GASTROENTERITIS | ENTERIC FEVER

Diarrhoea Fever

Peyer's patches
• Inflammation
• Ulceration

Peyer's patches
Lymphatic drainage
Bloodstream

Fever
PUO

4–5% ← Bacteraemia → 100%

Recirculation

Gallbladder
Peyer's patches
Ulceration

± varying
GI symptoms

Usual, spontaneous ← Recovery → 10% mortality
Long-term carriage rare   5% long-term carriage (↑ with age)

Symptomatic ← Treatment → Ciprofloxacin
± Ciprofloxacin
Treat bacteraemia

Resistance
↑ Worldwide
Need sensitivity testing

Salmonella

There is only a single species in the genus *Salmonella enterica* of which there are subtypes: *enterica*, *salamae*, *arizonae*, *diarizonae*, *houtenae*, and *Salmonella bongori*. The serotypes of *enterica* subspecies account for most human and warm-blooded animal infections.

Salmonellosis

Salmonella are host-adapted to animals, and infection in humans is usually confined to the bowel. Infection presents as acute self-limiting diarrhoea On some occasions the organisms are capable of causing invasive disease, including bacteraemia and life-threatening septicaemia or osteomyelitis.

The organisms are found in domestic animals. Human cases and convalescent carriers are also important sources. Transmission is faecal–oral, usually through ingestion of contaminated foods. Infection is more common and severe in patients with reduced gastric acid or the immunocompromised or splenectomized. It can be complicated by reactive arthritis or a chronic carrier state.

Enteric fever

Enteric fever (typhoid) is caused by *Salmonella enterica* serotype typhi or paratyphi. Invasion of the intestinal wall, with spread to local lymph nodes, is followed by primary bacteraemia and infection of the reticuloendothelial system. The bacteria reinvade the bloodstream and gut from the gallbladder, multiply in Peyer's patches, and cause ulceration that may be complicated by haemorrhage or perforation. Patients present with fever, alteration of bowel habit (diarrhoea or constipation) and the classical but rare rash (rose spots on the abdomen). Hepatosplenomegaly may also be demonstrated. Enteric fever may be complicated by osteomyelitis and, rarely, by meningitis.

Other infections

Urinary tract infection and pyelonephritis

Most *Escherichia coli* urinary tract infections are caused by a limited number of serotypes. These have some specialized adaptations including higher quantity of K antigen, adherence to uroepithelial cells via pili and haemolysin production. Mannose-resistant pili are associated with pyelonephritis. *Proteus* spp. also possess specialized adherence pili, mediating attachment to urinary epithelium. Urease production by *Proteus* is the most important virulence determinant in urinary infection, lowering pH and precipitating stone formation.

Meningitis and brain abscess

Escherichia coli is an important cause of neonatal meningitis and is associated with a high mortality. Strains often express copious amounts of K1 capsular antigen. Meningitis may also follow neurosurgical procedures, especially when prosthetic devices are inserted. Enterobacteriaceae are often found as part of the polymicrobial flora of brain abscess.

Osteomyelitis and septic arthritis

Osteomyelitis or septic arthritis caused by *Salmonella* is an important complication for patients with sickle cell disease or HIV. Bone and joint infections with *Salmonella* and other Enterobacteriaceae are also found in older patients. Infection with other Enterobacteriaceae can also follow penetrating trauma when contaminated fragments are taken into the bony tissue. Treatment often includes a fluoroquinolone such as ciprofloxacin which penetrates into bony tissue.

Klebsiella infections

Infection with *Klebsiella* spp. is usually acquired in a hospital environment. They are an important cause of ventilator-associated pneumonia, urinary tract infection, wound infection and bacteraemia. Outbreaks of infection in high-dependency patients are described and are associated with septicaemia and a high mortality. Primary pneumonia with *K. pneumoniae* subspecies *pneumoniae* is a rare, severe, community-acquired infection, associated with a poor outcome. *Klebsiella rhinoscleromatis* causes a progressive granulomatous infection of the nasal passages and surrounding mucous membranes. Most infections are found in the tropics. *Klebsiella ozanae* has been associated with chronic bronchiectasis.

Enterobacter, *Serratia* and *Citrobacter* infections

These are environmental organisms and may colonize and infect hospitalized patients, causing wound infections, bacteraemia and hospital-acquired pneumonia. Many isolates may be naturally resistant to antibiotics and treatment choices are limited.

Diagnosis

Enteric Gram-negative organisms are identified by biochemical reactions, such as the pattern of fermentation of different sugars. Epidemiological investigation uses serotyping (sera directed against the lipopolysaccharide (O) antigens and flagellar (H) antigens), phage typing or colicine typing (using the pattern of inhibition produced by these proteins). Modern molecular typing methods are now also used.

A diagnosis of typhoid is made by isolating organisms from the blood or bone marrow.

Treatment and prevention

Most enteric Gram-negative organisms are susceptible to aminoglycosides, extended-spectrum cephalosporins, fluoroquinolones, β-lactams and carbapenems (e.g. meropenem). As some produce β-lactamases and aminoglycoside-degrading enzymes, treatment should be guided by sensitivity tests. Emerging extended-spectrum β-lactamase-carrying strains increase the resistance to broad-spectrum antibiotics.

In urinary tract infections, cefalexin, ampicillin, nitrofurantoin or trimethoprim are the antibiotics of first choice.

Diarrhoeal disease can be avoided by good hygiene, food preparation and safe water supplies. Treatment is primarily by oral rehydration (see Chapter 51).

Ciprofloxacin is the treatment of choice for typhoid; alternatives are trimethoprim or third-generation cephalosporins. Multidrug-resistant typhoid has been a major problem in some countries. A live attenuated vaccine (Ty21A) or a subcellular vaccine (containing the Vi antigen) is available for travellers to areas of high risk, but gives only partial protection.

Vibrio, Campylobacter and *Helicobacter*

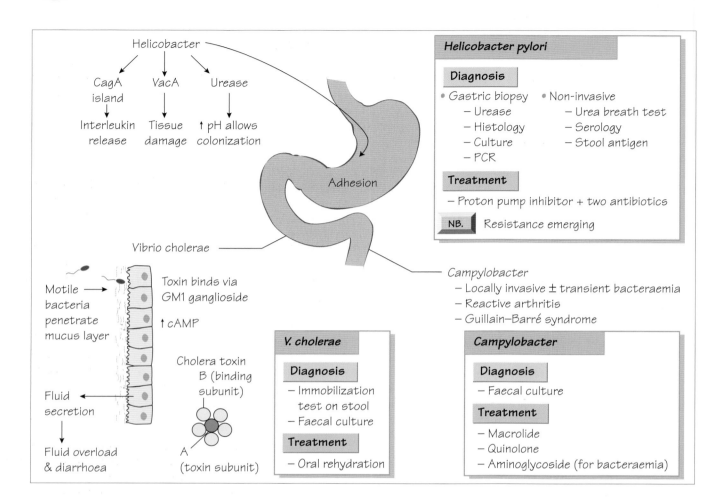

Vibrio spp.

Vibrios are small, Gram-negative, curved bacilli that are motile and oxidase-positive. There are more than eight species. *Vibrio cholerae* and *V. parahaemolyticus* are the main human pathogens.

Vibrio cholerae

The organism is subdivided by the somatic O antigens; only O1 and O139 are associated with cholera. It survives gastric acid, burrows through the intestinal mucus to attach to intestinal epithelial cells via the GM1 ganglioside and produces a multimeric protein toxin (cholera toxin). This stimulates adenyl cyclase within the cell, resulting in the secretion of water and electrolytes into the lumen of the bowel.

Epidemiology

Cholera is an exclusively human illness. The disease is transmitted via water and food, especially seafood. It is primarily a disease of developing countries where there is inadequate sanitation and lack of safe water supplies. Epidemic disease appears to follow a change in the environment, but is facilitated by human behaviour, such as war, refugee movements and migration. Cholera gives rise to periodic pandemics: the most recent epidemic is the seventh recorded.

Clinical features

Many infections with *V. cholerae* are mild or asymptomatic. The cholera syndrome is characterized by massive, painless, fluid diarrhoea, up to 20 L per day, which may be accompanied by vomiting. Severe dehydration follows, complicated by electrolyte imbalance.

Diagnosis

Where cholera is endemic, the diagnosis is based on clinical features. Immobilization of cholera bacteria in a diarrhoeal stool with specific antiserum yields a rapid diagnosis. The organism can be cultivated on selective medium (thiosulphate citrate bile-salt sucrose medium, or enriched in alkaline peptone water). Biochemical identification and serotyping should be performed to confirm the diagnosis.

Treatment

Oral rehydration solution (salt and glucose mix) is effective, although intravenous fluids may be required for severely ill patients. Antibiotics (e.g. tetracycline or ciprofloxacin), can shorten the duration and reduce the severity.

Prevention and control

Provision of safe water supplies and community education do much to prevent epidemic cholera. Several experimental live attenuated and subunit vaccines are under trial.

Campylobacter spp.

Campylobacter are microaerophilic, curved, Gram-negative rods; they are motile by virtue of polar flagella. They are associated with diarrhoeal disease and cause infection more commonly than do *Salmonella* and *Shigella*. Although there are more than 18 species of *Campylobacter*, *C. jejuni* is responsible for 90% of *Campylobacter* gastrointestinal infections. Infection follows ingestion of contaminated meat, poultry, unpasteurized milk or contaminated water. *Campylobacter coli* causes bacteraemia in immunocompromised patients.

Pathogenicity

Campylobacter jejuni invades and colonizes the mucosa of the small intestine. Antibodies to GM1 ganglioside are associated with Guillain–Barré syndrome.

Clinical features

Patients typically complain of influenza-like symptoms, crampy abdominal pain and diarrhoea which may be blood stained. In children, the pain may be sufficiently severe to suggest appendicitis or intussusception. A self-limiting bacteraemia is common. In some patients Guillain–Barré syndrome associated with ascending demyelination with motor and sensory deficits follows a few weeks after *Campylobacter* infection. Reactive arthritis may also occur.

Diagnosis

Faecal samples should be inoculated onto a *Campylobacter*-specific medium containing lysed blood and a mixture of antibiotics, and incubated at 42°C in a microaerophilic atmosphere. Identification of bacteria is based on their growth at 42°C, their characteristic microscopic morphology ('seagull wing') and expression of catalase and oxidase.

Treatment

Diarrhoea is often self-limiting but patients may be treated with erythromycin or fluoroquinolones. An aminoglycoside may be added for patients who have septicaemia.

Prevention and control

Prevention of campylobacteriosis depends on good animal husbandry and abattoir practices, and good food hygiene in shops, dairies and the home.

Helicobacter pylori

Helicobacter pylori is a Gram-negative, non-sporing, microaerophilic, spiral bacillus which is motile by virtue of five or six unipolar flagella. It is catalase-, oxidase- and strongly urease-positive.

Helicobacter cinaedi and *H. fennelliae* have been isolated from HIV-positive individuals with proctocolitis and bacteraemia.

Pathogenesis

Helicobacter pylori express urease, raising the pH in the surrounding locality, and thus protecting the bacterium from the effects of gastric acid. The CagA pathogenicity island encodes a type IV secretion system that injects the CagA protein into the host cytoplasm where subversion of a variety of cellular functions occurs, resulting in IL-8 secretion and inflammatory cell recruitment. VacA, a secreted protein which damages cells, is associated with severe disease.

Clinical features

Infection, whether acute or chronic, is often asymptomatic. Chronic infection often takes the form of a low-grade gastritis. There is a very strong association with both gastric and duodenal ulceration, and infection is associated with an increased risk of gastric cancer.

Diagnosis

Infection may be confirmed by gastric and duodenal biopsy at endoscopy. The biopsy is examined histologically, by microbiological culture and by NAAT. Demonstration of urease activity in the biopsy suggests the diagnosis. An infected patient given an oral dose of ^{13}C-labelled urea will excrete labelled carbon dioxide in the breath, which can be detected. The diagnosis can also be made by detecting antibodies in serum or stool by EIA.

Treatment

A combination of antibiotics and a proton pump inhibitor (triple therapy) appears to be most successful, e.g. amoxicillin, metronidazole and omeprazole. Other therapeutic variations are equally valid, but the ideal combination is not yet clear. Reinfection with *H. pylori* in adulthood is unusual.

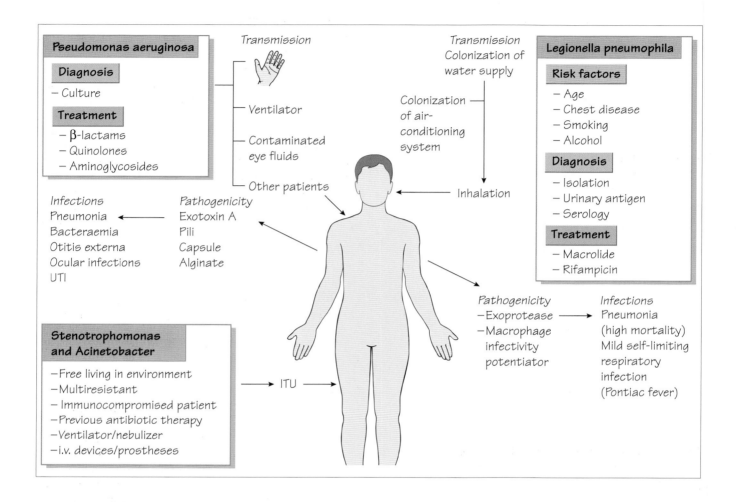

Pseudomonas spp.

The genus *Pseudomonas* is made up of environmental organisms that may cause opportunistic infections. Most cases are attributed to *Pseudomonas aeruginosa*.

Pseudomonas aeruginosa

This organism is a motile, aerobic, Gram-negative bacillus. It is ubiquitous in the environment, but rare in the flora of healthy individuals. Carriage increases with hospitalization. Moist environments harbour *P. aeruginosa*, such as sink-traps, taps and disinfectants in use for over 24 h.

Pathogenesis

Pseudomonas aeruginosa produces cytotoxins and proteases (e.g. exotoxins A and S, haemolysins and elastase). Isolates from patients with cystic fibrosis produce a polysaccharide alginate. This allows microcolonies to form where organisms are protected from opsonization, phagocytosis and antibiotics. The alginate, pili and outer membrane protein mediate adherence.

Alginate production is associated with hypersusceptibility to antibiotics, LPS deficiency, non-motility and reduced exotoxin production.

Clinical features

Corneal infection can be rapidly progressive, as can *Pseudomonas* otitis externa. Burns can become colonized, leading to secondary septicaemia. Septicaemia with a high mortality is a particular threat to neutropenic patients. A few show a destructive skin complication, ecthyma gangrenosum. Osteomyelitis, septic arthritis and meningitis can occur, the latter usually after neurosurgery. Chronic infection of cystic fibrosis patients causes a progressive deterioration of lung function.

Laboratory diagnosis

Pseudomonas aeruginosa grows on most media, but those containing cetrimide, irgasin and naladixic acid are selective. The organism is identified by biochemical testing and its ability to grow at 42°C. It may be typed by O and H agglutination reactions, bacteriophage typing, bacteriocin typing or molecular methods, such as pulse-field gel electrophoresis.

Treatment
Treatment is with aminoglycosides, carbapenems, ureidopenicillins, expanded-spectrum cephalosporins or fluoroquinolones. Organisms may exhibit multiple resistance.

Prevention and control
Vaccination is not effective. Spread of multiresistant strains should be controlled within hospitals by isolating infected individuals and reducing moist environments in which the organism may colonize.

Burkholderia spp.
Burkholderia cepacia
This organism causes chronic pulmonary infection among cystic fibrosis patients. It leads to a decline in pulmonary function or fulminant septicaemia. It spreads from person to person in cystic fibrosis clinics and is naturally resistant to many antibiotics. Treatment is with expanded-spectrum cephalosporins, carbapenems or ureidopenicillins, and is based on susceptibility testing.

Burkholderia pseudomallei
This is a free-living saprophyte of soil and water found in the tropics. It causes melioidosis, which presents as a tuberculosis-like disease, as acute septicaemia or as multiple abscesses. Septicaemia is associated with a high mortality. The diagnosis is made by cultivating the organism from blood or tissues. Treatment is with ceftazidime. *Burkholderia mallei* causes a similar infection in horses, known as glanders, which can spread to humans.

Stenotrophomonas maltophilia
Stenotrophomonas maltophilia is a Gram-negative bacillus normally found in soil and water. It inhabits moist environments and, because it is resistant to many antibiotics, can colonize patients in intensive-care units and the immunocompromised. Infection is transmitted by staff and by contaminated shared equipment, such as nebulizers. The organism causes septicaemia or pneumonia. Most strains are resistant to aminoglycosides and carbapenems but are sensitive to co-trimoxazole and tetracycline and sometimes to expanded-spectrum cephalosporins.

Acinetobacter spp.
Acinetobacter are small Gram-negative coccobacilli. They are environmental organisms that are naturally resistant to many antibiotics and colonize patients in hospitals, especially those in intensive-care units. They can colonize the inanimate environment in damp places, such as humidifiers, and are implicated in outbreaks of multidrug-resistant infection. Systemic invasion leads to pneumonia, septicaemia, meningitis or urinary tract infection. Infection is more likely in patients receiving antibiotics, patients with multiple cannulae and intubated patients. Treatment, when indicated, is based on the results of susceptibility tests.

Legionella spp.
Legionellae are fastidious, Gram-negative, pleomorphic bacteria. There are more than 39 species but *L. pneumophila* is most frequently implicated in human disease. *Legionella* spp. are found in rivers, lakes, warm springs, domestic water-supplies, fountains, air-conditioning systems, swimming pools and jacuzzis. The organisms multiply in water between 20 and 40°C, often in association with other microorganisms, such as cyanobacteria or *Acanthamoeba*. They are transmitted to humans when aerosols are generated and inhaled (e.g. in showers and air-conditioning systems). Legionnaires' disease is associated with previous lung disease, smoking and high alcohol intake, but previously healthy patients can be infected. Immunocompromised patients in hospital are vulnerable to infection if the hospital air-conditioning system is not adequately maintained.

Pathogenesis
Pathogenicity factors include the major outer membrane protein that inhibits acidification of the phagolysosome and the macrophage infectivity potentiator that is required for optimal internalization. *Legionella pneumophila* expresses a potent exoprotease.

Clinical features
Legionellosis may take the form of a mild influenza-like illness. Equally, pneumonia (Legionnaires' disease) can be severe with respiratory failure and high mortality. Patients often complain of gastrointestinal symptoms (e.g. nausea or vomiting and malaise) before lung symptoms become prominent. The cough is usually unproductive but dyspnoea is progressive. Psychiatric effects and confusion are common. Inappropriate naturetic hormone production may be associated with low serum sodium.

Laboratory diagnosis
Sputum, or preferably bronchoalveolar lavage fluid, should be cultured. Suspect colonies are identified serologically. Rapid diagnosis is by direct immunofluorescence or NAAT of respiratory specimens and antigen detection in urine. Serum antibodies rise after 10–14 days.

Treatment and prevention
Effective regimens usually consist of a macrolide antibiotic together with rifampicin.

Legionellosis is prevented by maintenance of air-conditioning systems and ensuring that hot-water supplies are above 45°C to prevent multiplication.

Chlamydia, Mycoplasma and Rickettsia

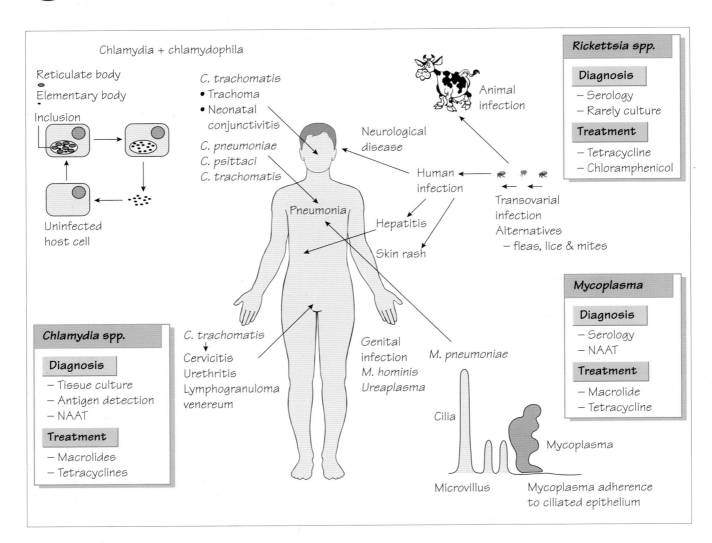

Chlamydia

There are three species: *C. trachomatis*, which infects the eye and the genital tract; and two respiratory pathogens, *C. psittaci* and *C. pneumoniae* (*Chlamydophila pneumoniae*). They are obligate, intracellular bacteria which exist in two forms: the reticulate body (non-infective intracellular vegetative form); and the elementary body (the extracellular form which permits the organism to survive and be transmitted) which is derived from the reticulate body by binary fission. All can cause pneumonia which may be severe, and conjunctivitis in different patient groups.

Pathogenicity

The major outer membrane protein may participate in attachment to mucosal cells. A 60-kDa cysteine-rich protein may also be associated with virulence.

Chlamydia psittaci

Chlamydia psittaci, a pathogen of birds and mammals, causes psittacosis. It is transmitted from birds to humans. After an incubation period of 10–14 days, fever, a dry unproductive cough, dyspnoea, headache and myalgia develop. Clinical examination reveals few signs of consolidation, but chest X-ray may show patchy consolidation.

Chlamydia pneumoniae

Chlamydia pneumoniae is transmitted from person to person by the respiratory route. It produces a pneumonia or bronchitis which is usually clinically mild but may be associated with pharyngitis, sinusitis and laryngitis. There is controversy over the role of *C. pneumoniae* in the development of atherosclerosis.

Chlamydia trachomatis

There are multiple serotypes. A–C are associated with trachoma (see Chapter 54) and neonatal conjunctivitis (see Chapter 43). D–K are associated with acute urethritis and pelvic inflammatory disease (see Chapter 49). Serotypes L1–L3 are associated with lymphogranuloma venereum.

Laboratory diagnosis

Psittacosis is usually diagnosed serologically using species-specific microimmunofluorescence, IgM tests and IgM capture EIA. *Bartonella* endocarditis may cause some cross-reactivity with chlamydiae in serological tests. Absorption or NAAT of heart valve tissue will differentiate these.

Chlamydia trachomatis is readily cultivated and antigen-detection EIA is available, but NAAT methods are now the diagnostic method of choice.

Chlamydia pneumoniae can be grown in HeLa cells. Complement fixation tests (CFTs), EIAs and NAAT-based methods are used for routine diagnosis.

Mycoplasma and *Ureaplasma*

Mycoplasma and *Ureaplasma* are small bacteria that lack a cell wall. They are parasites of animals, arthropods and plants. *Mycoplasma pneumoniae* is a primary human pathogen. *Mycoplasma hominis* and *U. urealyticum* are commensal, but may be associated with human infection of the genital tract.

Mycoplasma pneumoniae

This is an important cause of atypical pneumonia. After *Streptococcus pneumoniae*, it is probably the second most common cause of acute, community-acquired pneumonia.

Pathogenicity

Mycoplasma pneumoniae adheres to host cells by the P1 protein, a 169-kDa antigen. Immunity is short-lived: antigenic variation in the P1 protein is responsible for this. *Mycoplasma pneumoniae* locates itself at the base of the cilia where it induces ciliostasis. Secreted hydrogen peroxide damages host membranes and interferes with superoxide dismutase and catalase. Opsonized *M. pneumoniae* is readily killed by macrophages and by the activity of the complement system.

Clinical features

Patients present with fever, myalgia, pleuritic chest pain and a non-productive cough; headache is a prominent symptom. Antibodies that agglutinate the host's own red cells at low temperature result in peripheral and central cyanosis after exposure to the cold. Infection is associated with reactive (postinfective) arthritis, and neuritis.

Laboratory diagnosis

Culture growth of either *M. pneumoniae* or *U. urealyticum* is too slow to be of clinical value. A fourfold rise in CFT between acute and convalescent specimens indicates acute infection. An IgM-specific EIA, using a P1 protein surface antigen, is more sensitive than CFT, giving a positive result on a single specimen. NAAT is increasingly important in diagnosis.

Treatment

These organisms are resistant to β-lactams and cephalosporins, but are sensitive to erythromycin, tetracycline, aminoglycosides, rifampicin, chloramphenicol and quinolones.

Rickettsia

These organisms are obligate intracellular bacteria with biochemical similarities to Gram-negative bacteria. Clinically, they are divided into three groups:

1 spotted fever
2 scrub typhus
3 typhus.

Spotted fevers are transmitted by ticks. Rocky mountain spotted fever (RMSF) is caused by *Rickettsia rickettsii*. The typhus group includes *R. prowazekii* and *R. typhi* which cause epidemic and murine typhus, respectively; scrub typhus is caused by a single species, *R. tsutsugamushi*.

The incubation period is up to 14 days. After non-specific symptoms, patients may develop fever, arthralgia and malaise, followed by the development of a rash, conjunctivitis and pharyngitis. Confusion only occurs in a proportion of RMSF cases. Relapse of *R. prowazekii* infection months or years later is known as Brill–Zinsser disease and is usually milder than the primary infection.

Diagnosis is by immunofluorescence, CFT, IgM-specific EIA or NAAT. Tetracyclines and chloramphenicol are the treatment of choice, but must be initiated early to influence the outcome.

Coxiella burnetii

Coxiella burnetii is a small, Gram-negative, rod-shaped bacterium closely related to *Rickettsia*. It usually infects cattle, sheep and goats, localizing in the placenta. It survives desiccation in the environment and is transmitted predominantly by the aerosol route.

Coxiella burnetii is the causative organism of Q fever which may present as an atypical pneumonia or pyrexia of uncertain origin. In about 50% of infected patients the clinical picture is dominated by hepatitis and splenomegaly. Relapses take the form of culture-negative endocarditis or granulomatous hepatitis.

Q fever is usually diagnosed by CFT using acute and convalescent serum. *Coxiella* express different antigens at different phases of infection. EIA and NAAT-based methods are more sensitive and are positive earlier in the course of the disease.

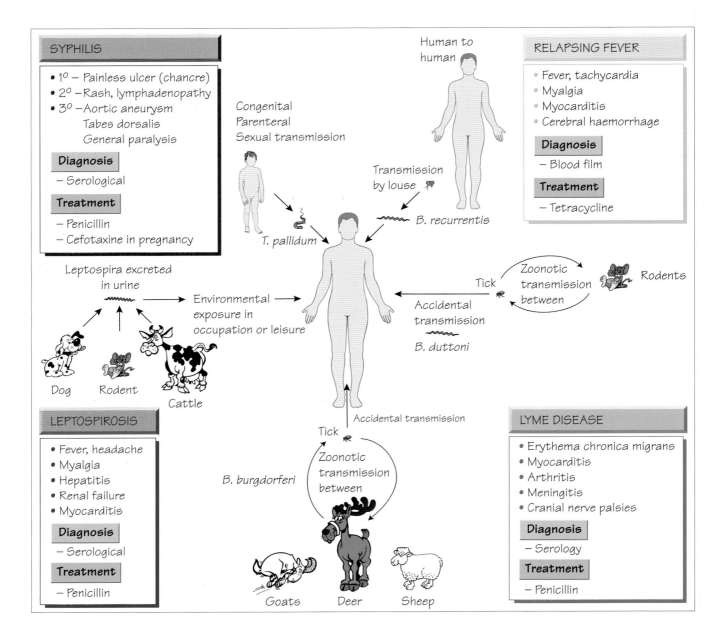

SYPHILIS
- 1° – Painless ulcer (chancre)
- 2° – Rash, lymphadenopathy
- 3° – Aortic aneurysm
  Tabes dorsalis
  General paralysis

Diagnosis
- Serological

Treatment
- Penicillin
- Cefotaxine in pregnancy

Congenital
Parenteral
Sexual transmission

T. pallidum

Human to human

Transmission by louse

B. recurrentis

RELAPSING FEVER
- Fever, tachycardia
- Myalgia
- Myocarditis
- Cerebral haemorrhage

Diagnosis
- Blood film

Treatment
- Tetracycline

Leptospira excreted in urine

Environmental exposure in occupation or leisure

Dog   Rodent
Cattle

Tick   Zoonotic transmission between   Rodents

Accidental transmission

B. duttoni

LEPTOSPIROSIS
- Fever, headache
- Myalgia
- Hepatitis
- Renal failure
- Myocarditis

Diagnosis
- Serological

Treatment
- Penicillin

B. burgdorferi

Tick
Zoonotic transmission between

Accidental transmission

Goats   Deer   Sheep

LYME DISEASE
- Erythema chronica migrans
- Myocarditis
- Arthritis
- Meningitis
- Cranial nerve palsies

Diagnosis
- Serology

Treatment
- Penicillin

Leptospira

Leptospira are aerobic, tightly coiled, motile bacteria. Previously there were thought to be two species: *L. interrogans*, and *L. biflexa*—a non-pathogen. These have now been reclassified as a single species, *Leptospira interrogans*, which has more than 200 serovariants. More than 17 genospecies have been described that include both pathogenic and non-pathogenic serovars. These serovars are conventionally described as if they were species.

Epidemiology

Leptospira are parasitic, with different preferred mammalian hosts, for example the rat is the reservoir of *L. icterohaemorrhagiae*. Leptospires colonize the renal tubules of their natural host and are excreted in the urine. Humans may be infected from contact with animal urine or by contaminated water or soil. Water sports enthusiasts and agricultural and abattoir workers are at increased risk of infection.

Pathogenesis and clinical features

The central nervous system, liver and kidneys are most affected in human disease. The severity varies between serovars; for example *L. icterohaemorrhagiae* infection is usually more severe than *L. copenhageni* infection.

Leptospirosis has two phases: first there is bacteraemia, with fever, headache, myalgia, conjunctivitis and abdominal pain; then, after the organisms disappear from the blood, fever, uveitis and aseptic meningitis predominate. Jaundice, haemorrhage, renal failure and myocarditis occur in severe cases and poor outcome is associated with hypotension, renal failure and clinical evidence of pulmonary involvement.

Diagnosis and treatment

Although *Leptospira* can be cultured from the blood during the first week of illness, rising antibody titres or IgM-specific EIA are the usual diagnostic methods. A WHO diagnostic scheme combines six clinical and two laboratory features to make a diagnosis.

Penicillin or doxycycline must be commenced early in the disease. Doxycycline is an effective prophylactic agent if exposure to infection is likely to have occurred. Killed whole-cell vaccines are used in some countries.

Borrelia

Borrelia are loosely coiled, spiral bacteria. They are transmitted to humans via arthropods: lice or ticks. Infections arise sporadically throughout the world, with a well-defined geographical territory and host specificity; for example, humans are the only host of louse-borne relapsing fever (*B. recurrentis*). Epidemics arise during war or mass migration when humans invade the *Borrelia*–tick–rodent habitat.

Relapsing fever

Borrelia invade the bloodstream, producing fever. Antibodies clear the organism from the blood, but antigenic variation allows relapse. Untreated, relapsing fever resolves when the organism exhausts its repertoire of antigenic variation.

Patients experience headaches, myalgia, tachycardia and rigors; examination may reveal hepatosplenomegaly and a petechial rash. Episodes last for 3–6 days; relapses are approximately a week apart. Louse-borne relapsing fever has a high mortality (up to 40%); tick-borne disease mortality rarely exceeds 5%. Dysrhythmias (secondary to myocarditis), cerebral haemorrhage or hepatic failure are the usual causes of death. Postexposure treatment with doxycycline has been shown to be effective in preventing disease.

Lyme disease

Borrelia burgdorferi and the related species *B. afzelii* and *B. garinii* cause Lyme disease, and are transmitted by *Ixodes* ticks. It is endemic in the eastern US and Europe, and humans are accidental hosts. Localized skin infection is followed by migration through the skin and dissemination throughout the body. The early symptoms are caused by the acute infective process; later manifestations are thought to be related to the host immune response.

Initially an expanding red macule or papule (erythema chronicum migrans) may be found. This is followed by headache, conjunctivitis, fever and regional lymphadenopathy. New skin lesions, myocarditis, arthritis, aseptic meningitis, cranial nerve palsies and radiculitis are complications. Acrodermatitis chronica atrophicans, a red skin lesion, may also occur. Diagnosis is by EIA and analysis of Western blot.

Vincent's angina

This is a painful, ulcerative, synergistic infection with *Borrelia vincenti* and fusobacteria or other anaerobes in the mouth. Clinical diagnosis is confirmed by Gram stain. Treatment is with penicillin and metronidazole.

Diagnosis and treatment

Diagnosis is by visualizing *Borrelia* in the peripheral blood. Lyme disease can be diagnosed using specific EIA.

Doxycycline or amoxicillin is used for treatment of early Lyme disease, and ceftriaxone for late or recurrent disease. Doxycycline is the drug of choice for treating relapsing fever. An outer membrane protein vaccine against Lyme disease shows promise in trials.

Treponema pallidum

The causative organism of syphilis, *Treponema pallidum*, is transmitted sexually and congenitally. The characteristic syphilitic lesions (gummae)—necrosis and obliterative endarteritis with fibroblastic proliferation and lymphocyte infiltration—are found throughout the body. The incidence is increasing worldwide, having fallen for many years.

Related organisms, *T. pertenue* and *T. carateum*, cause yaws and pinta, respectively. They are non-venereal and spread by contact, usually in childhood. Once common in the tropics, they are now rare as the result of an eradication campaign.

Clinical features

Organisms penetrate intact skin and then disseminate throughout the body. There are four phases of disease: primary chancre (painless ulcer with rubbery edge and regional lymphadenopathy); secondary (an acute febrile illness with a generalized non-itchy scaling rash, typically involving the palms, and associated with lymphadenopathy); a long latent phase lasting many years; and tertiary (systemic lesions become symptomatic, e.g. aortitis, posterior cord degeneration and dementia).

Diagnosis

Treponema pallidum can be seen by dark-ground microscopy in specimens from the primary chancre or rash, and NAAT can be used in this situation. Serology uses EIA for specific IgG and IgM and improved diagnostic accuracy of neonatal syphilis is obtained by detecting IgM by Western blotting. Alternative tests use cardiolipin agglutination (measures disease activity) and tests based on cultivated treponemes such as the treponemal haemagglutination test (TPHA). CSF testing should be performed to detect early central nervous system involvement.

Treatment

Syphilis is treated with penicillin (or either azithromycin or tetracyclines if the patient is allergic). An acute febrile response (the Jarisch–Herxheimer reaction) may develop in some patients after the first dose of antibiotics. Careful serological follow-up is required to confirm cure and/or detect early central nervous system involvement.

28 Virus structure, classification and antiviral therapy

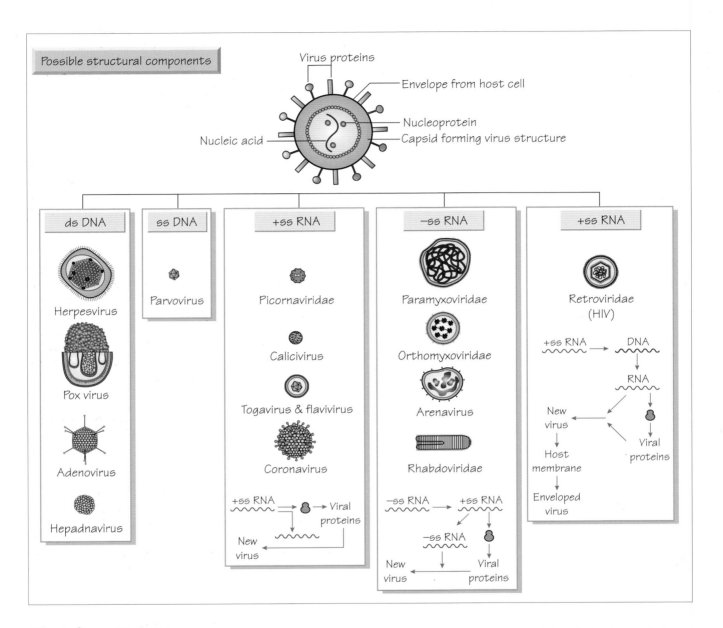

Viral classification

Modern classification of viruses is based on the genetic material, the mode of replication, the structure and symmetry of the structural proteins (capsids) and the presence or absence of an envelope.

Genetic material and replication

DNA viruses

Virus DNA is either double stranded (ds) or single stranded (ss). Double-stranded DNA viruses include poxviruses, herpesviruses, adenoviruses, papovaviruses and polyomaviruses. These last two are small viruses associated with benign tumours, such as warts, and malignant tumours, such as cervical cancer.

Hepatitis B virus is double stranded with single-stranded portions. Among the single-stranded DNA viruses, parvoviruses are responsible for erythema infectiosum.

DNA viruses usually replicate in the nucleus of host cells, producing a polymerase which reproduces viral DNA. Viral DNA is not usually incorporated into host chromosomal DNA.

RNA viruses

RNA viruses possess a single strand of RNA and adopt different reproduction strategies depending on whether the RNA is sense or antisense. RNA sense (positive) may serve directly as mRNA. It is translated into structural protein and an RNA-dependent

RNA polymerase. A virus with RNA antisense (negative) contains an RNA-dependent RNA polymerase that transcribes the viral genome into mRNA. Alternatively, the transcribed RNA can act as a template for further viral (antisense) RNA.

Retroviruses possess single-stranded sense RNA that cannot act as mRNA. It is transcribed into DNA by reverse transcriptase and incorporated into host DNA. The subsequent transcription to make mRNA and viral genomic RNA is under the control of host transcriptase enzymes.

Capsid symmetry

Viral nucleic acid is covered with a protein coat made up of repeating units (capsids), with either icosahedral or helical symmetry. In icosahedral symmetry, the capsids form an almost spherical structure. Helical symmetry is found in RNA viruses that have capsids bound around the helical nucleic acid. A structure based on repeating units reduces the number of genes devoted to the viral coat and simplifies the process of viral assembly.

Envelope

In some viruses the nucleic acid and capsid proteins of the virus (the nucleocapsid) are surrounded by a lipid envelope derived from the host cell or nuclear membranes. The host membrane is altered by viral-encoded proteins or glycoprotein, which may act as receptors for other host cells. Enveloped viruses are sensitive to substances that dissolve the lipid membrane (e.g. ether).

Antiviral therapy

The intracellular location of viruses and their use of host cell systems makes antiviral therapy difficult to develop. Despite this, there are a growing number available: some of the more important are listed below.

Amantadine

Amantadine prevents uncoating and release of viral RNA. Resistance arises readily. Short courses may prevent disease during outbreaks but are usually reserved for patients at high risk.

Nucleoside analogues

Aciclovir is phosphorylated by virally encoded thymidine kinase. This is not a human enzyme but only occurs in virally infected cells. As it is active against herpesviruses and varicella zoster virus, it is used to treat herpes simplex infections and for prophylaxis against herpes infections in the immunocompromised. Resistance occurs through the development of deficient thymidine kinase production or alteration in the viral polymerase gene. The drug can be taken orally and crosses the blood–brain barrier. Most is excreted unchanged in the urine; toxicity is rare.

Virus-infected cells phosphorylate **ganciclovir** to a monophosphate form, which is then metabolized further to a triphosphate form. The drug is active against herpes simplex virus and cytomegalovirus. It is indicated in the treatment of life- or sight-threatening cytomegalovirus infections in immunocompromised individuals. Ganciclovir causes bone-marrow toxicity (monitoring of haematological indices is required during therapy).

Lamivudine is a nucleoside inhibitor active against hepatitis B virus. There are several additional nucleoside and nucleotide inhibitors being developed as alternative treatments for this infection, including adefovir, entecavir, tenofovir, telbivudine and clevudine.

Ribavirin is a guanosine analogue that has activity against respiratory syncytial virus, influenza A and B, parainfluenza virus, Lassa fever, hantavirus and other arenaviruses. It inhibits several steps in viral replication including capping and elongation of viral mRNA. Its mechanism of action is probably by inhibition of cellular pathways. Usually, ribavirin is administered by aerosol for treatment of severe respiratory syncytial virus infection in infants.

Anti-retroviral compounds

Highly active anti-retroviral therapy (HAART) used in HIV infection brings about improvement in the CD4 count and a fall in the viral load, with a corresponding fall in the incidence of opportunistic infections. Survival and quality of life are improved. New agents are constantly coming into use.

Nucleoside reverse transcriptase inhibitors

Nucleoside reverse transcriptase inhibitors (NRTIs) are nucleoside analogues that inhibit the action of reverse transcriptase, the enzyme responsible for the conversion of viral RNA into a DNA copy. These include the longest established anti-retroviral drug zidovudine (AZT), lamivudine (3TC), stavudine (d4T), tenofivir, didanosine (ddI) zalcitabine (ddC) and abacavir. They are the mainstay of retroviral therapy and are used in combination during initial therapy (see Chapter 44).

Protease inhibitors

Protease inhibitors are the most effective anti-retroviral compounds available because when used as a single agent they produce the greatest fall in viral load. They include atazanivir, indinavir, lopinivir, ritonavir and saquinavir.

Non-nucleoside reverse transcriptase inhibitors

Non-nucleoside reverse transcriptase inhibitors (NNRTIs) such as nevirapine, efavirenz and delavirdine inhibit reverse transcriptase by an alternative mechanism to NRTIs. They have been shown to be effective agents in combination regimens. As resistance occurs after a single mutation, they are only used in maximally suppressive regimens.

New drugs

Enfuvirtide is a new drug that inhibits HIV fusion with cells. It is administered subcutaneously.

Inhibitors of viral uncoating

Pleconaril inhibits uncoating of picornaviruses and inhibits viral attachment to host cells. It is active against enteroviruses and rhinoviruses. It is absorbed orally and clinical trials suggest it shortens clinical symptoms.

Other new drugs

Ruprintrivir is a novel agent which selectively inhibits human rhinovirus 3C protease. It is administered by nasal spray and appears to have useful activity in rhinovirus infection.

29 Herpesviruses I

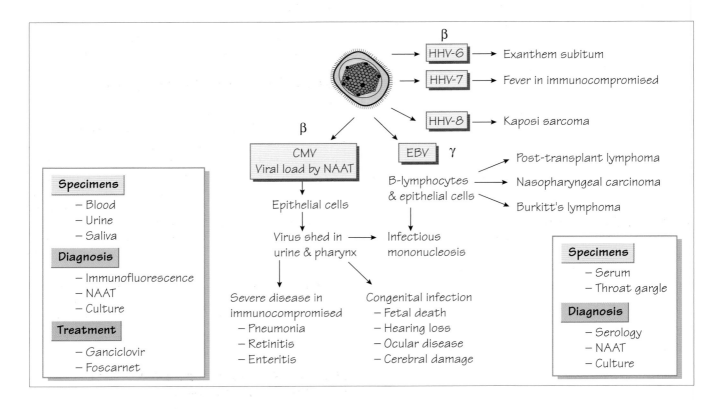

Herpesviruses

Herpesviruses are large, enveloped, double-stranded DNA viruses (120–200 nm). The genome is 120–240 kb, coding for more than 35 proteins. The envelope contains glycoproteins and Fc receptors. All infections are lifelong, with latency following the acute primary episode and relapses occurring later in life, especially if the individual becomes immunosuppressed.

Classification

Herpesviruses are divided into α-herpesviruses (fast-growing cytolytic viruses that establish latent infections in neurones, e.g. herpes simplex and varicella zoster); β-herpesviruses (slow-growing viruses that become latent in secretory glands and kidneys, e.g. cytomegalovirus, CMV); and γ-herpesviruses (latent in lymphoid tissues, e.g. Epstein–Barr virus, EBV).

The recently discovered human herpesvirus types 6 and 7 (HHV-6 and HHV-7) might be classified as γ-herpesviruses because they become latent in lymphocytes, but they have most genetic homology with γ-herpesviruses and so are classified with this group. The Kaposi sarcoma-associated virus, human herpesvirus type 8 (HHV-8), has homology with EBV. Herpesviruses are antigenically diverse; only herpes simplex (HSV) 1 and 2 share antigenic similarity.

Cytomegalovirus

CMV has a similar structure to other herpesviruses. Infection persists for life and virus is shed in the urine and saliva. Approximately 50% of adults in the UK have been infected.

Epidemiology and pathogenesis

Infection is transmitted vertically or from close person-to-person contact. Under poor socioeconomic conditions infection is acquired early, but with increasing wealth this is delayed. Pregnant women who develop infection may transmit it to the fetus before or after birth. Infection can also be acquired from blood transfusion or organ transplantation, where acute infection is associated with severe disease.

Clinical features

In congenital infection neonates may either be severely affected (see Chapter 43), or initially asymptomatic, later developing hearing defects or delay in developmental milestones. Postnatal infection is usually mild. Immunocompromised patients, especially those with organ transplantation or HIV, may develop severe pneumonitis, retinitis or gut infection, either through reactivation of latent virus or by acquisition from the donor organ.

Diagnosis

Congenital infection is confirmed by detecting virus in the urine within 3 weeks of birth. In adult infection, CMV can be cultured, or the virus detected by NAAT from specimens of urine or blood. Monitoring viral load is important in identifying patients with severe disease who require treatment.

Treatment and prevention

Severe life- or sight-threatening infection should be treated with ganciclovir, together with immunoglobulin in the case of pneumonitis. Valganciclovir, the ester of ganciclovir, is an oral preparation used for initial treatment and maintenance. Alternatives, all of which are more toxic, include foscarnet and cidofovir, a DNA polymerase chain inhibitor. Appropriate screening of donor organs and blood products can reduce the risk of transmission.

Epstein–Barr virus

The EBV genome codes for Epstein–Barr nuclear antigen complex (EBNA), latent membrane protein, terminal protein, the membrane antigen complex, the early antigen (EA) complex and the viral capsid antigen.

Epidemiology and pathogenesis

As with CMV, infection is generally found in the very young in developing countries and in adults in industrialized countries. Gaining entry via the pharynx, the virus infects B cells and disseminates widely. EBV is capable of immortalization of B cells. This can result in neoplasia: Burkitt's lymphoma (found in sub-Saharan Africa in association with malaria), nasopharyngeal carcinoma (in China) and lymphoma (in immunosuppressed patients). It is the cause of lymphoproliperative disease in transplant recipients.

Clinical features

Infection is characterized by fever, malaise, fatigue, sore throat, lymphadenopathy and occasionally hepatitis. It usually lasts about 2 weeks, but persistent symptoms may develop in a few patients. EBV infection is associated with tumours (see above).

Diagnosis

Diagnosis is made by a rapid slide agglutination technique. Definitive diagnosis is by detection of specific IgM to EBV viral capsid antigen. NAAT based diagnosis can now also be used.

Human herpesviruses 6 and 7

These β-herpesviruses were first isolated in the 1980s and are the sole members of the *Roseolovirus* genus. Herpesvirus 6 is divided into two subtypes, A and B, which infect human T cells. Transmission is probably through infected saliva; almost all individuals are infected by the end of their second year. Infection is associated with exanthem subitum and is characterized by 3–5 days of febrile illness which settles as the rash appears. However, infection without rash is common. HHV-6 is an important cause of febrile convulsions and encephalitis. The latter is a rare complication, as is hepatitis, sometimes fulminant. IgG EIA is available and NAAT (quantitative NAAT) may be helpful in diagnosis.

Infection with HHV-7 is almost universal by the age of 5. There is no clear association between infection with HHV-7 and clinical illness, but increasing association between the virus and encephalitis.

Diagnosis is with paired serum to detect antibody.

Human Kaposi sarcomavirus or human herpesvirus 8

HHV-8 is a γ-herpesvirus related to herpesvirus samiri found in simians. Transmission is vertical from mother to child, and in the young is by mucosal (non-sexual) contact. Initial infection is characterized by infectious mononucleosis syndrome. Later, immunocompromised patients, especially those with HIV, develop Kaposi sarcoma. Diagnosis is principally by NAAT from suspicious tissues. Serological tests using EIA and indirect fluorescence are available.

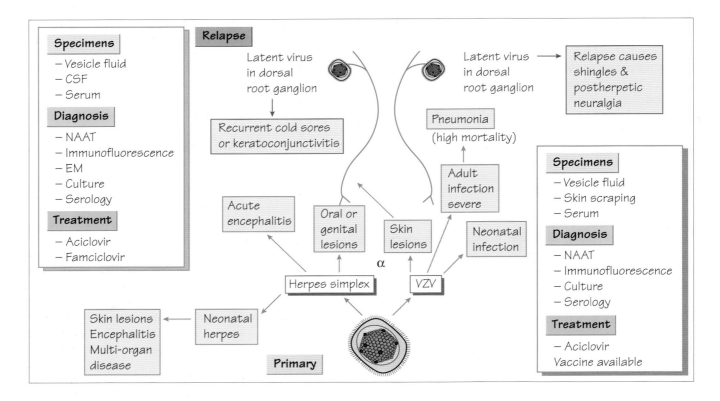

Herpes simplex
Pathogenesis and epidemiology

Herpes simplex is transmitted from person to person by direct contact. It produces skin vesicles because of its cytolytic activity. The virus may invade locally and, via the sensory neurones, remains latent in the sensory ganglia. Reactivation may be triggered by physical factors (e.g. infection, sunlight), or psychological stress. As cell-mediated immunity controls infection, patients with immunosuppressive conditions are at risk of reactivation and severe infection.

Clinical features

Herpes simplex virus 1 (HSV-1) infection is often asymptomatic, but young children commonly develop fever, vesicular gingivostomatitis and lymphadenopathy; adults may exhibit pharyngitis and tonsillitis. Primary eye infection produces severe keratoconjunctivitis: recurrent infection may result in corneal scarring. Primary skin infection (herpetic whitlow) usually occurs in traumatized skin (e.g. fingers). Fatal encephalitis may occur (see Chapter 47). Maternal transmission during childbirth may result in generalized neonatal infection and encephalitis.

HSV-2 infections cause painful genital ulceration that can be severe, with symptoms lasting up to 3 weeks. Recurrent infections are milder and virus shedding is short-lived, but infection can be transmitted to sexual partners during this time. Genital herpes is an important cofactor in the transmission of HIV. Meningitis is an uncommon complication of primary type 2 infection.

Diagnosis

NAAT detection of virus is increasingly the standard diagnostic method. HSV is readily grown in tissue culture from fresh specimens (e.g. vesicle fluid, and genital and mouth swabs) and the virus can be visualized by electron microscopy, in specimens of vesicle fluid.

The role of serology is limited. The ratio between serum and CSF antibody may indicate local production and can help in the diagnosis of HSV encephalitis. MRI or CT scans of the brain may detect temporal lobe lesions typical of herpes encephalitis.

Treatment

Topical, oral and intravenous preparations of antiviral agents, e.g. aciclovir, are available for the treatment of HSV infections. Agents with better oral absorption include valciclovir and famciclovir. Encephalitis is treated with intravenous aciclovir.

Varicella zoster virus

Varicella zoster virus (VZV) (125 kb) has only one serological type and causes the acute primary infection known as chickenpox or varicella, and its recurrence (shingles).

Pathogenesis and epidemiology

Vesicular fluid contains large numbers of VZV: when the vesicle ruptures, VZV is transmitted by airborne spread. The attack rate in non-immune individuals is very high (> 90%).

The incubation period is 14–21 days and the disease is most common in children aged 4–10 years. Individuals are infectious from a few days before the rash appears until the vesicle fluid has dispersed. Recovery provides lifelong immunity.

The virus lies latent in the posterior root ganglion and in 20% of those previously infected the virus will travel down the axon to produce reactivation lesions in that dermatome, known as 'shingles'. As lesions of shingles contain VZV they are infectious to the non-immune who may contract chickenpox; it is impossible to contract shingles directly from chickenpox.

Clinical features

Most of the discomfort of VZV infection arises as a result of the rash; systemic symptoms are mild. Individual lesions progress through macules and papules to vesicular eruptions which, following rupture, develop a crust and spontaneously heal. They appear in crops, usually 2 or 3 days apart, and affect all parts of the body including the oropharynx and genitourinary tract. The rash lasts for 7–10 days, but complete resolution may take as long again. Rarely, infection can be complicated by haemorrhagic skin lesions that can be life-threatening. Lesions can become secondarily infected with skin bacteria (e.g. *S. aureus* or *S. pyogenes*).

VZV virus pneumonia is more common in adults, especially immunocompromised individuals. It has a high mortality; survivors may recover completely or may have respiratory impairment. Although minor postinfectious encephalitis can also develop, fatal disease is rare. Maternal transmission, through contact with vaginal lesions during birth, can result in severe neonatal infection.

Shingles is a painful condition which is more common in elderly people. Ocular damage may follow the involvement of the ophthalmic division of the trigeminal nerve. Up to 10% of shingles episodes will be followed by postherpetic neuralgia, a condition which may last for many years and is associated with a significant risk of suicide.

Diagnosis

The diagnosis is usually made clinically. NAAT diagnosis of VZV is now the main method of diagnosis. Direct staining of vesicle fluid may reveal characteristic giant cells. VZV may be seen by electron microscopy and can be cultivated in tissue culture. Serology is used to determine the immune status of patients considered at risk (e.g. the immunocompromised or pregnant women) to reduce the risk of spread in institutional outbreaks.

Treatment and prevention

Antiviral agents (e.g. aciclovir) may be used for both adult chickenpox and shingles. The incidence of postherpetic neuralgia may be reduced by the use of valaciclovir or aciclovir. The pain may be severe and require referral to a pain clinic.

Primary infection can be prevented by a live attenuated virus vaccine but this is not currently part of the routine immunization schedule. It is not licensed for general use in many countries; however it is now available to healthcare workers in the UK. Zoster immune globulin can be administered to neonates and others at risk of serious disease. The fetus may be affected if primary infection occurs in early pregnancy. Exposed non-immune pregnant women should be screened and offered varicella immunoglobulin.

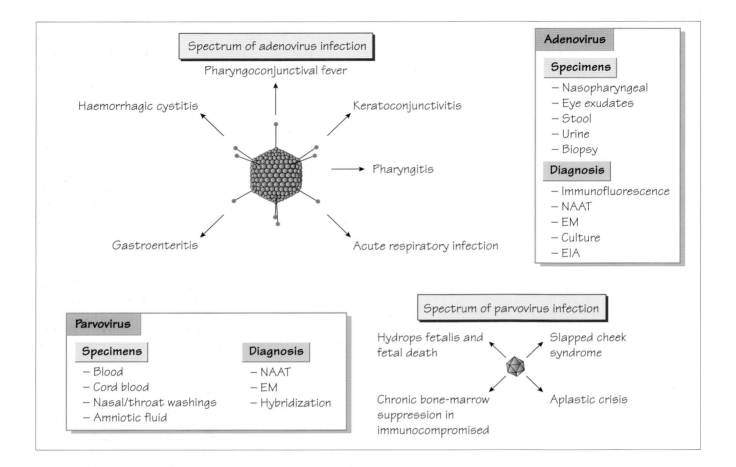

Adenovirus

Adenoviruses are unenveloped, icosahedral, double-stranded DNA viruses. The capsid is made up of 240 hexon and 12 vertex penton capsomeres with vertex fibre projections. The hexons possess a group-specific antigen and all human adenoviruses share a common hexon antigen. The pentons also share a group-specific antigen, but the fibres contain type-specific antigens. There are more than 50 serotypes of human adenoviruses divided into six groups (A–F) based on genomic homology.

Epidemiology and pathogenesis

Group A strains cause asymptomatic enteric infection, groups B and C cause respiratory disease, group D cause ketatoconjunctivitis, group E cause conjunctivitis and respiratory disease, and group F cause infantile diarrhoea. Infection shows little seasonal variation and is usually transmitted by the faecal–oral or respiratory routes. Eye infections, however, can be transmitted by hand–eye contact, a process facilitated in swimming pools; epidemic keratoconjunctivitis is highly infectious. Genetically modified adenoviruses are increasingly being explored as vectors for gene therapy.

Clinical features

Clinical syndromes include acute febrile pharyngitis (serotypes 1–7), pharyngoconjunctival fever (3, 7) acute respiratory infection and pneumonia (1, 2, 3, 7), pertussis-like syndrome (5) conjunctivitis (3, 7), epidemic keratoconjunctivitis (8, 19, 37), enteritis (40, 41) and acute haemorrhagic cystitis (11, 21). Immunocompromised patients may suffer severe pneumonia (1–7), urethritis (37) and hepatitis in liver allografts.

Diagnosis

Virus can be isolated in tissue culture from stool specimens and rectal, conjunctival or throat swabs. Serotype is determined by using specific neutralizing antibody or haemagglutination inhibition, or NAAT. Enteric adenovirus can be visualized by electron microscopy or detected by EIA or NAAT. A fourfold rise in complement-fixing group-specific antibody indicates recent infection.

Prevention and control

Live attenuated vaccines were available for types 4 and 7 and used by US military recruits. Swimming pool outbreaks of ocular infection are prevented by adequate chlorination. Single use and adequate decontamination of equipment and appropriate hygiene by healthcare staff prevents transmission between patients undergoing ophthalmic examination.

Parvovirus

Parvoviruses are small, unenveloped, icosahedral, single-stranded DNA viruses (5.6 kb, 8–26 nm). There is only one parvovirus, B19, that is known to cause human disease, and it is given the genus name *Erythrovirus*.

Epidemiology

Infection is found worldwide and throughout the year. Transmission is by the respiratory route; outbreaks of erythema infectiosum may occur in schools. Seroprevalence increases with age with more than 60% of adults possessing antibody.

Pathogenesis and clinical features

Parvovirus B19 invades red cells through the globoside P antigen which is responsible for this tissue tropism. It replicates in dividing cells and thus targets immature erythrocytes. Clinically, it produces erythema infectiosum, a mild febrile disease, typically in young children who may exhibit a 'slapped cheek' appearance. A symmetrical small joint arthritis may develop, especially in adults. Arrest of red cell production is well tolerated in healthy individuals but causes a worsening of anaemia in patients with a high red blood cell turnover (e.g. aplastic crises in patients with sickle cell disease). The risk of infection in pregnancy is low but it may lead to hydrops fetalis and fetal death. There is no evidence that parvovirus causes congenital abnormalities but infection during the second trimester results in spontaneous abortion in 10% of pregnancies.

Diagnosis

This is usually clinical but NAAT is the test of choice as cultivation is difficult. Detection of IgM is also used. Blood, nasal or throat washings, cord blood and amniotic fluid can be examined by electron microscopy.

Prevention and control

No vaccine is available at present. Respiratory precautions should prevent transmission in the hospital environment.

Papillomavirus

These are small, enveloped, double-stranded DNA viruses with more than 100 types. Some are responsible for common warts and genital warts. Types 16 and 18 predominate in cervical neoplasia, They are transmitted by close contact, including the sexual route. Diagnosis of common wart is clinical, cervical neoplasm is diagnosed by cytology and NAAT. A vaccine against types 16 and 18 is entering clinical use.

Poxvirus

Poxviruses are double-stranded DNA viruses with complex symmetry and a shape resembling a ball of wool.

Smallpox

Smallpox was once a major cause of death worldwide. The World Health Organization coordinated an international campaign of vaccination which resulted in the eradication of the disease in 1977. There are concerns that smallpox may become a bioterrorism weapon, which have prompted some countries to produce stocks of vaccine.

Monkeypox

This zoonotic organism is found in remote villages in the rainforest of central Africa. The clinical features are similar to smallpox. The disease is severe and the fatality rate is greater than 10% in unvaccinated subjects. Transmission between humans does not occur readily. There have been outbreaks traced to the importation of exotic animals into non-endemic countries. Diagnosis is by electron microscopy or NAAT.

Orf

This organism causes cutaneous pustular dermatitis in sheep and goats and can be transmitted to humans. Infection is characterized by a single vesicular lesion, typically on the finger, which resolves spontaneously after a few weeks. Although the virus can be cultivated, diagnosis is usually made by clinical appearance and a history of exposure.

Molluscum contagiosum

This is a common condition, especially in children. There are crops of small regular papular skin lesions, usually on the face, arms, buttocks and back. The pearl-like surface may exhibit a central dimple. It may be transmitted sexually or by direct contact. Topical steroids or HIV make the extent of disease greater.

Microscopically, the papules are caused by epidermal hypertrophy which extends into the dermis. Cells with inclusion bodies are seen in the prickle-cell layer. The diagnosis is usually made clinically and can be confirmed by electron microscope examination of scrapings from the lesion.

Lesions may continue to appear for up to 1 year in immunocompetent individuals. Molluscum may be a chronic problem for HIV-positive individuals requiring cidofovir ointment. Traditional treatment—by prodding the lesions with a sharp implement—promotes healing.

Tanapox

Tanapox is a febrile illness with usually a single nodular skin lesion that may ulcerate and which heals spontaneously. Infection is acquired in central and east Africa; the diagnosis is usually suggested by travel history and can be confirmed by electron microscopy or NAAT.

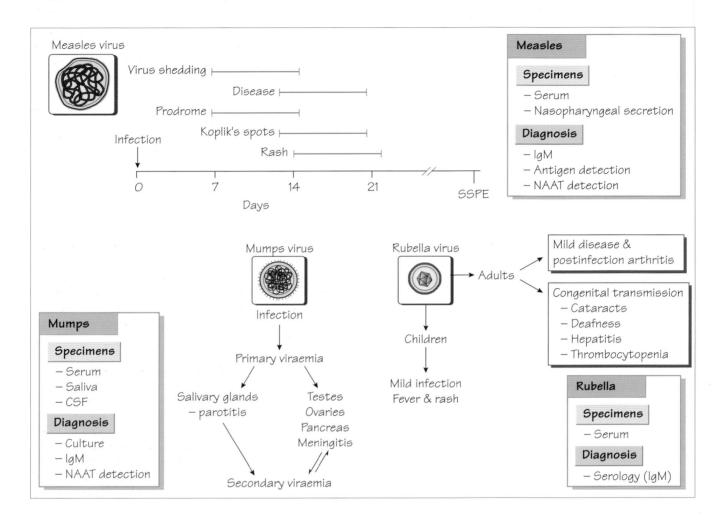

Measles

Measles, a negative single-stranded enveloped RNA virus, is a member of the genus *Morbillivirus* of the Paramyxoviridae family. There is a single serotype. The virus encodes six structural proteins, including two transmembrane glycoproteins, fusion (F) and haemagglutinin (H), that facilitate attachment to the host cell and viral entry. Antibodies to F and H are protective.

Pathogenesis and epidemiology

Measles virus initially infects the epithelial cells of the upper respiratory tract. Invasion of neighbouring lymphoid tissue leads to primary viraemia and the involvement of the reticuloendothelial system. Invasion of these cells leads to secondary viraemia and dissemination throughout the body, coincident with clinical symptoms.

Measles is transmitted by the aerosol route, with a high attack rate. The incubation period is 9–12 days. Children are infectious for 3 days before the rash emerges. Natural infection is followed by lifelong immunity. In healthy children mortality is rare except in the immunocompromised (e.g. HIV) or in malnutrition, especially vitamin A deficiency. The mortality is highest in children under 2 years of age. A population of 500 000 is required to sustain endemic spread. Recent falls in vaccine coverage have resulted in a reappearance of the disease.

Clinical features

Infection begins with a 2–4-day coryzal illness, when small white papules (Koplik's spots) are found on the buccal mucosa near the first premolars. The morbilliform rash, appearing first behind the ears, spreads centrifugally. After 3–4 days it changes to a brownish colour, often accompanied by desquamation. Secondary pneumonia, otitis media and croup are common complications. Acute postinfectious encephalitis occurs in approximately 1 in 1000 and is associated with a high mortality and morbidity. Subacute encephalitis, a chronic progressive disease, occurs mainly in children with leukaemia. Subacute sclerosing panencephalitis (SSPE) is a rare progressive fatal encephalitis that develops more than 6 years after infection.

Diagnosis

The diagnosis is made clinically but infection in vaccine recipients may be atypical. Laboratory confirmation is by serology using haemagglutination inhibition titres or an IgM-specific EIA. Salivary IgM detection has also proved valuable. SSPE is diagnosed by detecting virus-specific antibody synthesis in CSF (e.g. specific IgM). Diagnosis by NAAT (reverse transcriptase NAAT) and molecular characterization of the virus can be achieved.

Mumps

A member of the *Paramyxovirus* genus, the mumps virus is a pleomorphic, enveloped, antisense RNA virus. Mumps has one serotype.

Epidemiology

Infection usually occurs in childhood but many adults are susceptible as it has a lower attack rate than other childhood exanthemas. The incubation period is 14–24 days. Subclinical infection is common, especially in children. Infection is transmitted readily by the respiratory route. The virus invades the salivary glands, testes, ovaries, central nervous system and pancreas. Natural infection is followed by lifelong immunity. Epidemics re-emerge if vaccination coverage falls.

Clinical features

Mumps is characterized by fever, malaise, myalgia and parotid gland inflammation. Meningitis occurs in up to 15% of patients with parotitis; mumps virus was once one of the most common causes of viral meningitis. Complete recovery is almost invariable, although rare fatal forms and postmeningitis deafness may occur. Other complications, including orchitis (20%), oophoritis (5%) or pancreatitis (5%), are more common in adolescents and young adults, often after the parotitis has resolved.

Diagnosis

Mumps rarely needs laboratory diagnosis but this is required for public health reasons. Specific IgM in serum or saliva are diagnostic. Mumps virus can be isolated from saliva or detected by NAAT.

Rubella

Rubella, a rubivirus and member of the Togaviridae, is an icosohedral, pleomorphic, enveloped, positive-strand RNA virus. There is only one antigenic type of rubella virus.

Epidemiology

Endemic infection is rare in countries with efficient vaccine programmes. Transmission is via the respiratory route. Virus is shed from 7 days before until 14 days after the appearance of the rash. Maternal infection may cause congenital abnormalities in approximately 60% of cases, and the risk is highest during the first trimester. Natural infection is followed by solid immunity.

Clinical features

Rubella is associated with fever, a fine red maculopapular rash and lymphadenopathy. During the prodrome red pinpoint lesions occur on the soft palate. Arthritis (more common in females) and self-limiting encephalitis are complications.

Congenital infection is associated with fetal death or severe abnormalities, such as deafness, central nervous system deficit, cataract, neonatal purpura and cardiac defects.

Diagnosis

Diagnosis in pregnancy or pregnancy contact is required. Diagnosis can be made by detecting IgM and IgG antibody in serum or saliva. Congenital disease is confirmed by demonstrating the presence of specific IgM persistent antibodies (> 6 months) in an infant, or viral isolation from the infant after birth. NAAT (reverse transcriptase NAAT) and sequencing can be used for diagnosis and molecular epidemiology.

Prevention of measles, mumps and rubella

These three diseases can be prevented by a live attenuated combined vaccine (MMR). Vaccine is given between 13 and 15 months, with a booster dose given at school entry. Booster doses of measles vaccine may be required. The rapid antibody response to measles vaccine can be used to protect susceptible individuals exposed to measles. Women who are sexually active should be screened for the presence of rubella antibodies (e.g. when they first attend for contraceptive advice). Seronegative women should be offered rubella vaccination. Recent unproven concerns about the safety of MMR resulted in a fall in vaccine coverage and have resulted in a return of measles and mumps in many small outbreaks.

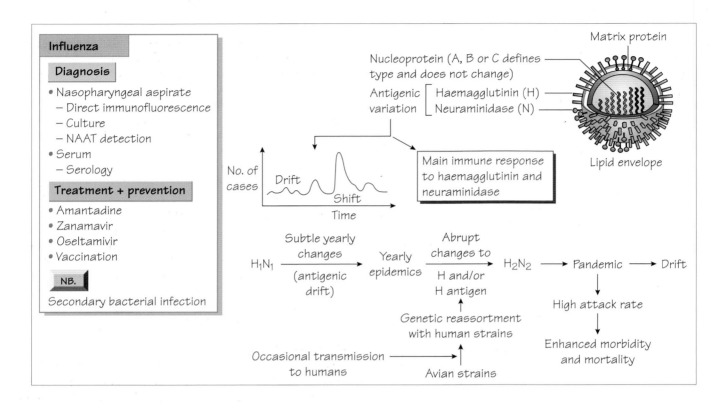

Influenza virus

Influenza virus is an enveloped orthomyxovirus (100 nm) containing a negative single-stranded RNA genome divided into eight segments. This facilitates genetic reassortment, allowing the development of different viral surface antigens. The virus expresses seven proteins, three of which are responsible for RNA transcription. The nucleoprotein has three antigenic types designating the three main virus groups, influenza A, B and C. Of the three, influenza A, and rarely B, undergo genetic shift. The matrix protein forms a shell under the lipid envelope with haemagglutinin and neuraminidase proteins expressed as 10-nm spikes on the envelope. These interact with host cells. Virus is typed based on the haemagglutinin (H) and neuraminidase (N) antigens.

Epidemiology

Annual epidemics of respiratory infection occur because of minor antigenic changes (antigenic drift). When there is a major antigenic shift, a worldwide pandemic may develop. Pandemics occur every 10–40 years, often originating in the Far East then circulating to the west. Such strains can often be traced to infected birds, poultry or pigs. Pandemic influenza A strains have a high attack rate and are associated with increased morbidity and mortality; 20 million people died in the 'Spanish flu' epidemic of 1919. The risk of a pandemic is high when there are epizootics of avian flu circulating in domestic birds (e.g.

H5N1). Avian strains can spread to humans and may be associated with a high mortality but not spread more widely. Mutation may permit person-to-person spead, triggering a pandemic. Serotypes B and C are exclusively human pathogens that do not cause pandemics.

There is a 1–4-day incubation period, with patients infectious for a day preceding and the first 3 days of symptoms. Headache, myalgia and fever and cough last for 3–4 days. Complications, more common in elderly people and patients with cardiopulmonary disease, include primary viral or secondary bacterial pneumonia.

Diagnosis

Diagnosis is usually clinical but rapid diagnosis by direct immunofluorescence and NAAT (reverse transcriptase NAAT) is used for vulnerable patients or to exclude/diagnose avian influenza. Laboratory diagnosis is also required for vaccine design where virus isolation is required. This process is coordinated internationally by the WHO. Serology provides a retrospective diagnosis.

Treatment, prevention and control

Treatment is usually symptomatic; secondary bacterial infections require appropriate antibiotics. Inactivated viral vaccines are prepared from the currently circulating viruses each year. Vaccination provides 70% protection and is recommended for individuals at risk of severe disease, such as those with cardiopulmonary disease or asthma. Influenza A infection can be prevented or modified in susceptible individuals with amantadine. The neuraminidase inhibitors zanamavir and oseltamivir shorten the duration of symptoms. They are indicated for patients who are at risk of severe complications and may have value in slowing the progression of a pandemic and reducing the associated mortality.

Parainfluenza virus

This is a fragile enveloped paramyxovirus (150–300 nm) containing a single strand of negative-sense RNA (15 kb). It has four types that share antigenic determinants.

Pathogenesis and epidemiology

The virus attaches to host cells where the envelope fuses with the host cell membrane. The virus multiplies throughout the tracheobronchial tree. Infection, transmitted by the respiratory route, peaks in the winter, with the highest attack rates in children under 3 years.

Clinical features

In this common, self-limiting condition, usually lasting 4–5 days, children are distressed, coryzal and febrile. In young children, hoarse coughing often alternates with hoarse crying and is associated with inspiratory stridor secondary to laryngeal obstruction (croup). Rarely, bronchiolitis, bronchopneumonia or acute epiglottitis may develop, signalled by reduced air entry and cyanosis.

Diagnosis and treatment

Diagnosis is clinical. Direct immunofluorescence gives rapid results and viral isolation and NAAT (reverse transcriptase NAAT) are available as part of a respiratory virus screen. Treatment is symptomatic (e.g. paracetamol and humidification). Severe infection can be treated with ribavirin and humidified oxygen.

Respiratory syncytial virus

This enveloped paramyxovirus (120–300 nm) containing a single strand of negative-sense RNA attaches to host cells by 12-nm glycoprotein spikes. There is antigenic variation within the two types, A and B.

Epidemiology

Respiratory syncytial virus (RSV) is found worldwide, infecting children during the first 3 years of life. There are yearly epidemics in the winter months in temperate countries and in the rainy season in tropical countries. RSV spreads readily in the hospital environment. The frail elderly and patients with compromised respiratory tract can develop serious infection.

Clinical features

After 4–5 days' incubation, coryza develops. Bronchitis in older children, and bronchiolitis in the very young, develops in 40%. Severe disease develops quickly but, with intensive care, mortality is very low. Children with bronchiolitis are febrile and tachypnoeic, with chest hyperinflation, wheezing and crepitations. Cyanosis is rare. Radiological appearances are variable and include hyperinflation and increased peribronchial markings.

Diagnosis and treatment

Direct immunofluorescence or EIA of nasopharyngeal secretions is rapid. Many laboratories use NAAT (reverse transcriptase NAAT) for diagnosis. The virus can be cultivated.

The treatment for RSV infection is based on symptomatic relief and humidification. Severe cases may require hospitalization and humidified oxygen. Severely ill immunocompromised patients may benefit from aerosolized ribavirin.

Prevention

No vaccine is currently available.

Coronavirus

This is a spherical enveloped virus (80–160 nm) with positive-sense linear ssRNA (27 kb); the envelope contains widely spaced club-shaped spikes. Coronaviruses cause a coryza-like illness similar to that of rhinovirus. The virus has been observed in the faeces of patients with diarrhoeal disease and asymptomatic subjects. Diagnosis is by serology using CFT or EIA, detection of coronavirus-specific antigens, or electron microscopy.

A virus emerged in China and was associated with severe pneumonia (SARS). It was transmitted by the respiratory and oral route, and mortality was approximately 10% but higher in elderly people and immunocompromised. Healthcare workers were vulnerable to infection, so stringent precautions are required to prevent hospital transmission. Coordinated infection control has permitted eradication of the virus.

Metapneumovirus

Human metapneumovirus, a paramyxovirus, has recently been identified from children with acute respiratory tract infections. It makes up just under 10% of cases occurring in winter months, with a clinical syndrome similar to RSV infection. Dual infection with RSV is associated with severe disease. Diagnosis is by NAAT (reverse transcriptase NAAT).

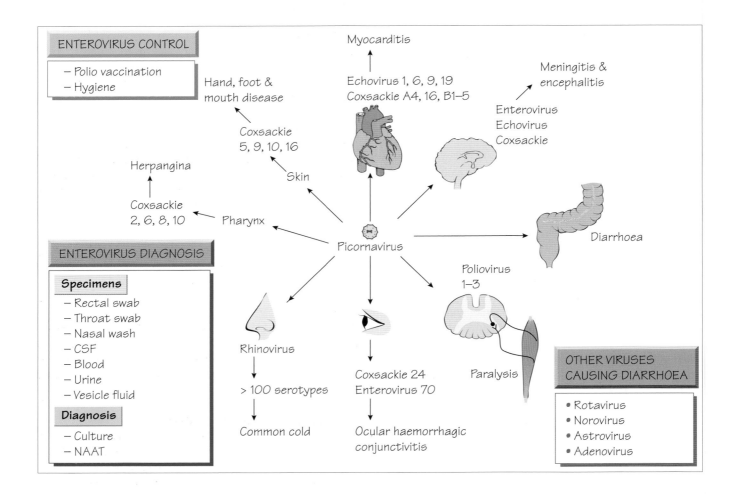

ENTEROVIRUS CONTROL
- Polio vaccination
- Hygiene

Hand, foot & mouth disease

Myocarditis

Echovirus 1, 6, 9, 19
Coxsackie A4, 16, B1–5

Meningitis & encephalitis

Enterovirus
Echovirus
Coxsackie

Coxsackie 5, 9, 10, 16

Skin

Herpangina

Coxsackie 2, 6, 8, 10

Pharynx

Picornavirus

Diarrhoea

ENTEROVIRUS DIAGNOSIS

Specimens
- Rectal swab
- Throat swab
- Nasal wash
- CSF
- Blood
- Urine
- Vesicle fluid

Diagnosis
- Culture
- NAAT

Rhinovirus

> 100 serotypes

Common cold

Poliovirus 1–3

Coxsackie 24
Enterovirus 70

Paralysis

Ocular haemorrhagic conjunctivitis

OTHER VIRUSES CAUSING DIARRHOEA
- Rotavirus
- Norovirus
- Astrovirus
- Adenovirus

Enterovirus

Enteroviruses and rhinoviruses belong to the Picornaviridae. Enteroviruses fall into three main serotype groups: poliovirus, coxsackievirus and enteric cytopathic human orphan (ECHO) virus (echovirus). Later viral serotypes have been given a number (e.g. enterovirus 68–72).

The viruses are unenveloped, icosohedral, symmetrical particles with single-strand positive RNA coding for four proteins, VP1–4. Antibodies derived from intact virus are type-specific but those derived from empty capsids cross-react with other enteroviruses.

Pathogenesis

Enteroviruses attach and enter cells by a specific receptor that differs for different virus types; receptor differences may influence tissue tropism. Enteroviruses usually enter the body by the intestinal tract. This is followed by a local viraemia and invasion of susceptible reticuloendothelial cells. A subsequent major viraemia leads to invasion of target organs (e.g. meninges, spinal cord, brain or myocardium). The poliovirus appears to spread along nerve fibres. If significant multiplication occurs within the dorsal root ganglia the nerve fibre may die, with resultant motor paralysis.

Epidemiology

Enteroviruses spread by the faecal–oral route. In developing countries infection occurs early in life; it occurs later in industrialized countries. Cases of polio have been documented in parents and carers of infants receiving live vaccine. Adult non-immunity may be because of the absence of a polio vaccination programme (e.g. in immigrant populations). In the tropics infection occurs throughout the year but in temperate countries it peaks during the summer.

Clinical features

Polio may present as a minor illness (abortive polio), aseptic meningitis (non-paralytic polio), paralytic polio with lower motor neurone damage and paralysis, or progressive postpoliomyelitis muscle atrophy (a late recrudescence of muscle wasting, sometimes decades after the initial paralytic polio). In paralytic polio, muscle involvement is maximal within a few days after commencement of paralysis; recovery may occur within 6 months.

Self-limiting aseptic meningitis is a common presentation of enterovirus infection (see Chapter 47), although severe focal encephalitis or general infection may present in neonates. Herpangina is a self-limiting painful vesicular infection of the pharynx caused by some coxsackievirus types. Coxsackie B viruses are important causes of acute myocarditis (see Chapter 46). Hand, foot and mouth disease is characterized by a vesicular rash of the palms, mouth and soles that heals without crusting.

Diagnosis and treatment

Most enteroviruses are readily cultivated in tissue culture. In patients with signs of meningitis, a specimen of CSF, a throat swab and faecal specimen should be sent for viral culture, although reverse transcriptase NAAT is used increasingly. With a multiplicity of enterovirus types, serological diagnosis is impractical.

The mainstay of management remains supportive care, although pleconaril shows benefit in the treatment of enteroviral meningitis. Artificial ventilation may be required in the case of polio.

Prevention

Polio can be prevented by vaccination but the efficiency is dependent on an adequate population uptake. Two vaccines are available: the oral live attenuated Sabin; and the killed parenteral Salk vaccine. The Sabin vaccine is used in the international polio eradication campaign. It provides rapid immunity and displaces the wild-type virus circulating in communities. The campaign is nearing success, with polio restricted to a few geographical locations. The Sabin vaccine is contraindicated in immunocompromised individuals, in whom the Salk vaccine is used.

Rhinovirus

The rhinovirus is responsible for the common cold, there are more than 100 serological types. During the short incubation period (2–4 days), the virus infects the upper respiratory tract, invading only mucosa and submucosa. Viral excretion is highest during symptoms; transmission is by close contact with infected individuals. Headache, nasal discharge, upper respiratory tract inflammation and fever may be followed by secondary bacterial infections such as otitis media and sinusitis. Infection occurs worldwide with a peak incidence in the autumn and winter. Immunity after infection is poor because of the multiplicity of serotypes. Ruprintrivir given by nasal spray shortens symptoms in clinical trials. A vaccine is impractical.

Rotavirus

Rotaviruses are unenveloped spherical viruses in a double capsid shell. They contain 11 double-stranded RNA segments coding for nine structural proteins and several core proteins. Rotaviruses possess group, subgroup and serotype antigens.

Pathogenesis

Rotaviruses infect small intestinal enterocytes; damaged cells are sloughed into the lumen, releasing viruses. Diarrhoea is caused by poor sodium and glucose absorption by the immature cells that replace the damaged enterocytes.

Epidemiology

Rotavirus infection is the most important cause of viral diarrhoea, most commonly in children between 6 months and 2 years of age. Morbidity is highest in the young. There are seasonal peaks in the winter in temperate countries. Antibody to the virus does not confer immunity to further infection.

Diagnosis

Laboratory diagnosis by NAAT (reverse transcriptase NAAT) is most sensitive and antigen can be detected by EIA or the viruses visualized by electron microscopy.

Treatment and prevention

Treatment is symptomatic and supportive. The risk of infection can be reduced by provision of adequate sanitation. Vaccines based on rotaviruses that normally infect animals configured to carry the relevant coat proteins of human rotaviruses are nearing regulatory approval.

Norovirus and astrovirus

Norovirus are caliciviruses that are an important cause of acute diarrhoea and vomiting in hospitals, care homes, cruise liners and other confined communities. They can be divided into three genogroups. Astroviruses are small spherical particles; more than five serotypes have been recognized.

Virus replication occurs in the mucosal epithelium of the small intestine, resulting in broadening and flattening of the villi and crypt cell hyperplasia.

These viruses usually cause a mild self-limiting acute diarrhoeal disease, but can present with sudden-onset projectile vomiting and explosive diarrhoea. Infection is transmitted by the faecal–oral route with symptoms developing after a short incubation period (24–48 h). Sudden outbreaks of norovirus infection may occur in institutions, requiring the units to close to new admissions. Outbreaks can be troublesome to control if new susceptible patients or staff come into the infected unit. Diagnosis is best made by NAAT (reverse transcriptase NAAT) as EIA and electron microscopy are less sensitive. Sequencing is required for epidemiological purposes and to monitor the design of NAAT detection assays.

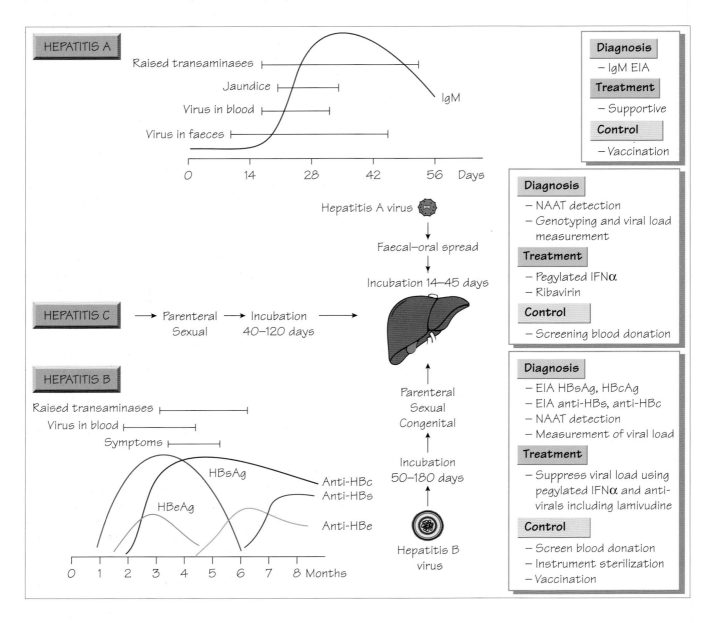

HEPATITIS A

Raised transaminases

Jaundice

Virus in blood

Virus in faeces

IgM

0   14   28   42   56   Days

Hepatitis A virus

Faecal–oral spread

Incubation 14–45 days

HEPATITIS C → Parenteral → Incubation →
Sexual   40–120 days

HEPATITIS B

Raised transaminases

Virus in blood

Symptoms

HBsAg

HBeAg

Anti-HBc
Anti-HBs

Anti-HBe

0 1 2 3 4 5 6 7 8 Months

Parenteral
Sexual
Congenital

Incubation
50–180 days

Hepatitis B
virus

Diagnosis
– IgM EIA
Treatment
– Supportive
Control
– Vaccination

Diagnosis
– NAAT detection
– Genotyping and viral load
  measurement
Treatment
– Pegylated IFNα
– Ribavirin
Control
– Screening blood donation

Diagnosis
– EIA HBsAg, HBcAg
– EIA anti-HBs, anti-HBc
– NAAT detection
– Measurement of viral load
Treatment
– Suppress viral load using
  pegylated IFNα and anti-
  virals including lamivudine
Control
– Screen blood donation
– Instrument sterilization
– Vaccination

Hepatitis may be caused by hepatitis viruses, other viruses (e.g. CMV, herpes simplex, EBV) and leptospirosis, and also has non-infective causes.

Hepatitis A

Hepatitis A virus (HAV) is a *Hepatovirus* related to *Enterovirus* in the family Picornaviridae. It has one serotype. The genome is a single strand of positive-sense RNA and there are four genotypes.

Transmission is faecal–oral and associated with summer, institutional outbreaks and point-source outbreaks following faecal contamination of water or food (e.g. oysters). Seroprevalence is highest in individuals in lower socioeconomic groups.

Anicteric infection is more common in the young, the risk of symptomatic disease increasing with age. Infection is characterized by a flu-like illness followed by jaundice, though some may not develop jaundice. Most patients make an uneventful recovery. Virus is shed in stool before jaundice appears.

Diagnosis

Anti-HAV IgM is diagnostic, appears before jaundice develops and persists for 3 months. IgG antibodies can be used to determine immune status. HAV RNA can be detected in the blood and stool during the acute phase of infection by NAAT, and sequencing can determine the relatedness of isolates for epidemiological investigations.

Treatment and prevention

Symptomatic treatment and support are all that are usually necessary. Chronic hepatitis does not occur.

Adequate sanitation and good personal hygiene will reduce the transmission of HAV. An effective inactivated vaccine is available for active protection. Individuals can be passively protected using human immunoglobulin.

Hepatitis B

Hepatitis B (HBV), a hepadnavirus, is a small enveloped virus containing 3.2-kb partially double-stranded DNA which codes for three surface proteins, i.e. surface antigen (HBsAg), core antigen (HBcAg), pre-core protein (HBeAg), a large active polymerase protein and transactivator protein. HBV is transmitted by parenteral, congenital and sexual routes.

Clinical features

There is a long incubation period (up to 6 months) before the insidious development of acute hepatitis, ranging from mild to severe. Fulminant disease carries a 1–2% mortality and 10% of patients develop chronic hepatitis complicated by cirrhosis or hepatocellular carcinoma. Congenital infection brings a high risk of hepatocellular carcinoma.

Diagnosis

Immunoassays for HBsAg, HBeAg, HBcAg and associated antibodies enable the diagnosis of acute infection and previous exposure. Viral load can be measured by NAAT and sequencing for resistance mutation allows monitoring of therapy and direction of drug choice.

Treatment and prevention

Interferon-alpha (α-interferon) treatment has limited efficacy but pegylated interferon (IFN) is superior for obtaining sustained suppression of HBV without drug resistance. Lamivudine, a nucleoside inhibitor, reduces viral load but resistance can emerge. New drugs, adefovir, entecavir, tenofovir, telbivudine and clevudine have equal or superior antiviral efficacy and can be used against lamivudine-resistant HBV. Antiviral suppression of HBV replication for 2–5 years reverses hepatic fibrosis, prevents cirrhosis and, when cirrhosis is established, improves liver function, prevents hepatic decompensation and lowers the risk of liver cancer. HBeAg seroconversion is often taken as a mark of treatment success.

Those at high risk should be immunized with recombinant HBV vaccine. Administration of vaccine and specific immunoglobulin to neonates of infected mothers reduces transmission. Blood donations are screened. Needle exchange programmes for drug misusers and sexual-health education schemes are beneficial.

Hepatitis C

Hepatitis C (HCV) is a positive-stranded RNA virus encoding a single polypeptide. Infection is mainly transmitted through infected blood. Seroprevalence is ~1% in healthy blood donors, higher in developing countries and highest in high-risk groups, such as those receiving unscreened transfusions. Healthcare workers are at risk. Sexual transmission and vertical transmission do occur but are uncommon.

Clinical features

Infection may cause a mild acute hepatitis but many cases are asymptomatic; fulminant disease is rare. HCV infection persists in up to 80% of patients; up to 35% of these develop cirrhosis, liver failure and hepatocellular carcinoma between 10 and 30 years later. This occurs because frequent virus mutation results in immunologically distinct 'quasi-species', allowing the organism to escape immunological control.

Diagnosis

HCV cannot be cultured; diagnostic antibody may be detected by EIA. NAAT (real-time) permits a rapid diagnosis. Sequencing of the virus defines the likelihood of response to therapy. Treatment can be monitored by measuring HCV RNA viral load.

Treatment and prevention

Ribavirin and pegylated α-interferon should be offered to all treatment-naïve patients. The response is best in patients with genotypes 1 and 2 and lowest initial viral load, but up to 80% will clear the virus. Liver fibrosis or necrotic inflammation from HCV infection is a major indication for liver transplantation.

Similar measures to those employed against HBV will also prevent transmission of HCV. There is no vaccine.

Hepatitis D

This defective RNA virus is surrounded by an HBsAg envelope. It is transmitted by intimate contact or by blood products and produces disease after a short incubation period, either as a coinfection with HBV or as a superinfection in an HBV carrier. Although asymptomatic infection may occur, hepatitis D (HDV) is associated with severe hepatitis and an accelerated progression to carcinoma. NAAT (real-time) is the most rapid method of making the diagnosis but antigen detection or IgM antibody detection by EIA can also be confirmatory. Preventive measures for HBV also protect against HDV.

Hepatitis E

This virus is a small, single-strand, non-enveloped RNA virus, classified in a separate genus: hepatitis E-like viruses. It is transmitted by the faecal–oral route, and outbreaks may occur after contamination of water supplies. Large outbreaks have occured in Asia. The diagnosis is made by detecting specific IgM or by real-time NAAT. Infection is prevented by hygiene measures.

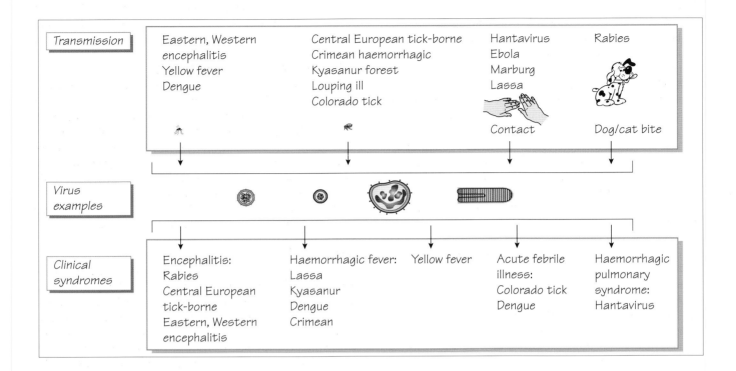

More than 100 viral infections cause encephalitis or haemorrhagic fever. Almost all are zoonoses, where the human is an accidental host that has come into contact with the natural life-cycle. They are transmitted by direct contact with blood and body fluids or by the bite of arthropods, such as mosquitoes, ticks and sandflies. Some infections are associated with a high mortality.

Rabies

Rabies is an acute rhabdovirus infection that, once symptoms develop, causes a fatal encephalomyelitis.

It is caused by a bullet-shaped RNA rhabdovirus with a helical nucleocapsid, enclosed in a membrane with 6–7-nm glycoprotein spikes and a single strand of negative-sense RNA. It infects warm-blooded animals worldwide. The virus is found in the saliva and it is transmitted to humans through the bite of an infected animal. Two epidemiological patterns exist: urban rabies, transmission in feral and domestic dogs; and sylvatic rabies, endemic in small carnivores in the countryside. Dog-bites are responsible for most infections. Bats, raccoons and skunks are an important reservoir and vector of infection in the Americas, whereas the red fox is the major reservoir of infection in Europe.

The virus gains entry to the nervous system via the motor endplates, spreading up the axons to enter the brain. Sites with short neural connections to the central nervous system have the shortest incubation period (7 days), whereas a bite on the foot may have an incubation period of 100 days. The depth of the bite and the concentration of viral inoculum also influence the incubation period.

A prodromal fever, nausea and vomiting precede the characteristic features of disease which takes one of two forms: furious rabies (hyperexcitability, hyper-reactivity, hydrophobia) or dumb rabies (an ascending paralysis). Disease is progressive and inevitably fatal. Diagnosis is based on the clinical features, confirmed by specific fluorescence in corneal scrapings, brain biopsy, or finding specific rabies antibody.

Disease may be prevented by appropriate wound care, local antiserum, systemic hyperimmunoglobulin and a postexposure vaccination course with the human diploid cell vaccine. Pre-exposure vaccination is reserved for high-risk groups – travellers to remote regions of endemic countries.

Yellow fever

Yellow fever virus is a flavivirus, an enveloped RNA virus with a single molecule of single-stranded RNA. It is transmitted to humans by the bite of an *Aedes aegypti* mosquito. Yellow fever is a zoonosis in which humans are an accidental host (sylvatic disease), but an urban cycle results in periodic human epidemics. Infection may be asymptomatic or cause an acute hepatitis and death resulting from necrotic lesions in the liver and kidney. After a short incubation period there is fever, nausea and vomiting followed later by jaundice. Haemorrhagic manifestations may develop and the vomitus may be black with digested blood (vomito negro). The mortality rate is high, but patients who recover do so completely. Diagnosis is confirmed by viral culture and serology. Disease prevention is by mosquito control and vaccination with the live attenuated vaccine.

Dengue

This *Aedes* mosquito-transmitted flavivirus is related to yellow fever virus. It has four serotypes. The incubation period is 2–15 days. A viraemia is present at the onset of fever and persists for several days. Dengue virus is found throughout the tropics and the Middle East. Epidemics occur when a new serotype enters the community or a large number of susceptible individuals move into an endemic area. Urban epidemics can be explosive and severe.

Following the sudden onset of fever and chills, headache and malaise, patients complain of pains in the bones and joints. Fever may be biphasic and a mild rash may also be present. Dengue haemorrhagic syndrome is a more severe form of the disease, with severe shock and bleeding diathesis. The mortality rate is 5–10%.

Dengue infection may be confirmed by serological means, culture and NAAT-based amplification methods. Dengue can only be prevented by controlling mosquito populations. Treatment is symptomatic.

Japanese B encephalitis

This is a mosquito-borne flavivirus infection that causes encephalitis with a high mortality. Pigs are the natural reservoir of the infection. Patients present with abrupt-onset fever and severe headache, nausea and vomiting. Convulsions can occur. There may be permanent cranial nerve or pyramidal tract damage. The infection may be prevented by vaccination.

West Nile virus

This virus has similar transmission and clinical features to dengue and Japanese B encephalitis. West Nile virus (WNV) was first detected in North America in 1999 and has spread across the continent into Canada, Latin America and the Caribbean.

Lassa fever

Lassa fever is a severe haemorrhagic fever caused by an arenavirus. Infection is transmitted from the reservoir house rat to humans, and from person to person by contact. The virus can affect all organs. Patients may present with fever, mouth ulcers, myalgia and haemorrhagic rash. Diagnosis is based on clinical symptoms and exposure history. Laboratory confirmation is by NAAT (reverse transcriptase NAAT) or serology. Ribavirin is effective if given early in the course of disease and can be given as postexposure prophylaxis to contacts.

Ebola and Marburg virus

These filovirus infections are found in Africa and are transmitted to humans from primates or from a rodent reservoir. They cause an acute haemorrhagic disease with high fever and mortality. Infection is transmitted by close contact, especially in the hospital environment. Treatment is supportive and benefit may be obtained from hyperimmune serum. Community control is not possible as the reservoir is not confirmed. Strict blood and body fluid precautions will prevent transmission in hospital. A vaccine using vesicular stomatitis virus encoding Marberg antigens can protect primates even if given shortly after infection.

Hantavirus

This bunyavirus infection is transmitted to humans from rodents and causes a haemorrhagic fever with renal failure or hantavirus pulmonary syndrome. The disease occurs widely throughout the world. Person-to-person spread does not appear to take place. The incubation period is 2–3 weeks, followed by fever, headache, backache and injected conjunctiva and palate. Hypotension, shock and oliguric renal failure follow. The mortality rate is about 5%.

Diagnosis of hantavirus infection is with serology tests and with reverse transcriptase NAAT. Ribavirin appears to improve mortality but controlled clinical trials have not been performed.

Nipah virus

Nipah virus, a paramyxovirus, can cause severe disease in both humans and animals. First described in Malaysia it is also found in other Asian countries, It typically causes febrile encephalitis with a high mortality rate. The reservoir is believed to be fruit bats, with infection acquired from contact with bats or an intermediate animal host such as pigs. Person-to-person spread occurs.

The related rarer Hendra virus is also acquired from bats and causes an influenza-like syndrome or encephalitis.

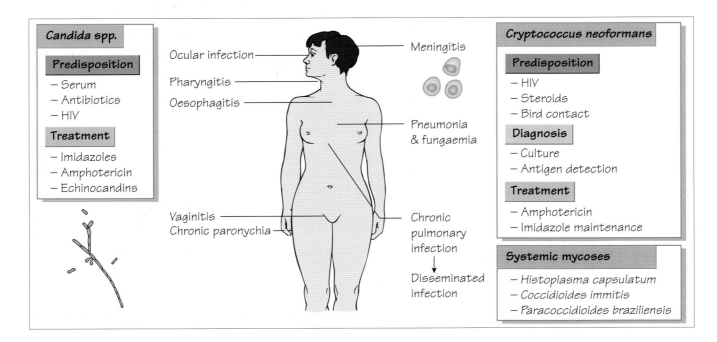

Fungi cause a wide range of diseases, ranging from cutaneous dermatophyte infections to invasive infection in the severely immunocompromised patient. They may have a yeast-like morphology (see below), or be filamentous (see Chapter 38).

Candida spp.

Candida spp. are widely distributed in the environment. They form part of the normal commensal population of the skin, the gastrointestinal tract and the female genital tract. Following the use of broad-spectrum antibacterials, fungal overgrowth may develop into infection. Patients with immunodeficiencies are particularly susceptible to this progression. Most infections are caused by *C. albicans*. Infection with other species such as *C. tropicalis, C. parapsilosis, C. glabrata* and *C. pseudotropicalis* are a problem in immunocompromised patients because they may be resistant to the antifungal agents used in therapy or prophylaxis.

Pathogenesis

Although these organisms have adhesins and extracellular lipases and proteinases, they have modest capacity to invade. Infection occurs when the natural resistance provided by the normal bacterial flora is altered by antibiotics, or where there is a severe loss of immune function.

Clinical features

Candida spp. cause pain and itching with creamy curd-like plaques on mucosal surfaces which bleed when removed. Skin and nailbed infections are common. In the immunocompromised patient, pharyngitis and oesophagitis can be severe; the associated dysphagia may lead to weight loss, an AIDS-defining illness. Systemic invasion is common in neutropenic patients. *Candida* spp. may also cause systemic and line-associated infection following broad-spectrum antimicrobial therapy in intensive-care patients.

Laboratory diagnosis

As *Candida* spp. form part of the normal flora, the significance of individual isolates can only be determined in relation to the overall clinical picture. *Candida* may be visualized microscopically and it grows readily on simple laboratory media. Molecular detection is of increasing importance for diagnosis in immunocompromised patients. Species identification is by biochemical techniques or sequencing of the 18S rRNA gene.

Antifungal susceptibility

Candida spp. are susceptible to amphotericin, with the exception of *C. lusitaniae*. They are usually susceptible to the imidazoles (e.g. fluconazole) and to 5-flucytosine. Caspofungin is useful in candidaemia.

Cryptococcus neoformans

Cryptococcus neoformans is the only species of this genus that regularly causes infection in humans. It is a saprophyte and animal commensal; the composition of pigeon faeces favours its growth. It is a rare cause of chronic lymphocytic meningitis in patients with lymphoma, those taking steroid or cytotoxic therapy and those with intense exposure, such as pigeon fanciers.

Cryptococcus is recognized as an important pathogen in T-cell-deficient patients.

Pathogenesis

The pathogenicity depends on an antiphagocytic capsule, melanin production and several lytic enzymes.

Clinical features

Infection usually presents as subacute meningitis, although pneumonia and fungaemic shock are recognized. In AIDS patients, relapses are common and lifelong suppressive therapy is necessary.

Laboratory diagnosis

It may be directly visualized in CSF by Gram stain or India ink. A latex test can detect capsular polysaccharide antigen. The organism may be isolated on blood or Sabouraud's agar; it is identified by biochemical tests or 18S rRNA sequencing.

Treatment

Amphotericin is the treatment of choice; liposomal preparations may be used to reduce toxicity. Flucytosine and fluconazole may also be used.

Pityriasis versicolor

Malassezia furfur infects the stratum corneum, causing brown, scaly macules. Patients with AIDS may develop severe dermatitis. Topical application of antifungal agents is usually successful.

Systemic yeast infections

Five main species are associated with systemic infection: *Histoplasma capsulatum*, *Histoplasma capsulatum* var. *duboisii*, *Blastomyces dermatitidis*, *Coccidioides immitis* and *Paracoccidioides brasiliensis*.

Infection is acquired by the respiratory route. They have a defined geographical distribution: south-west USA, South America and Africa. Severe disease is more likely in patients with reduced cell-mediated immunity.

Clinical features

Although usually asymptomatic or self-limiting, pulmonary or cutaneous infection may disseminate in infants or the immunocompromised, causing severe illness.

Laboratory diagnosis

These infections are diagnosed by microscopy and culture of blood sputum, CSF, urine or pus. The organisms are hazardous, and should be handled in a specialized containment facility.

Treatment

Patients with severe disease may be treated with amphotericin B.

Antifungal compounds

Azoles

The azole group of compounds (clotrimazole, miconazole, fluconazole and itraconazole) act by blocking the action of cytochrome P450 and sterol 14α-demethylase. This latter enzyme allows the incorporation of 14-methyl sterols in the fungal membrane, instead of ergosterol. Resistance can develop during long-term treatment.

Clotrimazole and miconazole are frequently used as topical preparations for minor infections.

Fluconazole

Fluconazole can be given orally, topically and parenterally. It is widely distributed, crosses the blood–brain barrier and is active against *Candida* and *Cryptococcus* but not filamentous fungi. It is used for the prophylaxis and treatment of cryptococcal infections and treatment of superficial and systemic candidiasis. Although well tolerated, it may cause liver enzyme abnormalities. It has significant drug interactions, increasing the serum concentration of phenytoin, cyclosporin and oral hypoglycaemic agents and reducing the rate of warfarin metabolism.

Itraconazole

In addition to being effective against *Candida*, *Cryptococcus neoformans* and *Histoplasma*, itraconazole also displays activity against filamentous fungi, including *Aspergillus* and the dermatophytes. It is indicated in treatment of invasive candidiasis, cryptococcosis, aspergillosis, superficial mycoses and pityriasis versicolor. Resistance is rare. It is well absorbed and can be given orally, achieving high tissue concentrations.

Voriconazole and posoconazole

Voriconazole is a broad-spectrum triazole that is active against many yeasts and moulds including *Aspergillus*. It has been reported to have a better success rate in proven invasive *Aspergillus* infection than amphotericin, but treatment is associated with transient visual disturbance. Posoconazole has a wide spectrum of activity. Further agents are in development.

Flucytosine

This synthetic fluorinated pyrimidine inhibits *Candida* spp., *Cryptococcus neoformans* and some moulds. The drug disrupts protein synthesis. It is well absorbed orally and can be given intravenously. Bone-marrow suppression, thrombocytopenia and abnormal liver function tests are adverse events. Resistance develops rapidly with monotherapy.

Echinocandins

Caspofungin was the first echinocandin. These act by inhibiting the synthesis of 1,3-β-glucan, a homopolysaccharide in the cell wall of many pathogenic fungi. They are active against both *Candida* and *Aspergillus*.

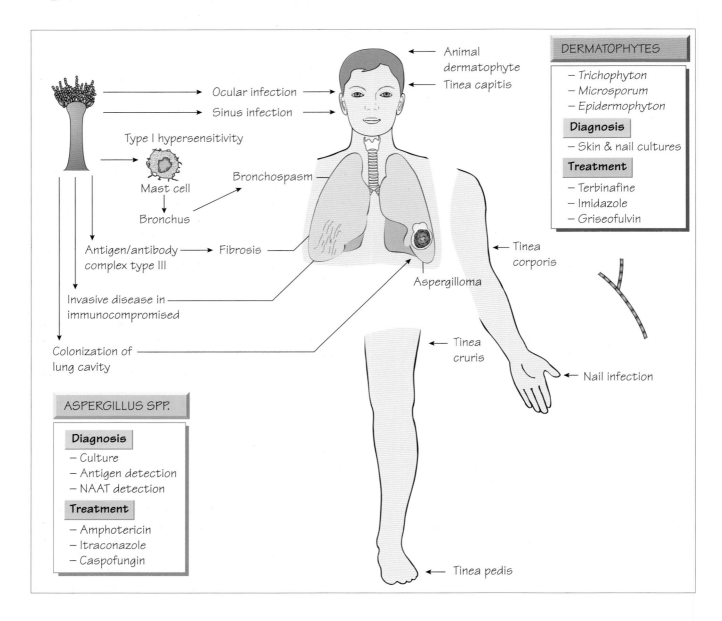

Aspergillus spp.

Aspergillus spp. are ubiquitous, free-living, saprophytic organisms. Four are regularly associated with human infection: *A. fumigatus*, *A. niger*, *A. flavus* and *A. terreus*.

Clinical features

Inhalation of *Aspergillus* spores may give rise to a type III hypersensitivity reaction with fever, dyspnoea and progressive lung fibrosis (farmer's lung). Some patients colonized by *Aspergillus* develop a type I hypersensitivity reaction resulting in intermittent airway obstruction (bronchopulmonary allergic aspergillosis).

Healed cavities or old bronchiectasis can become colonized with *Aspergillus*, an aspergilloma, or 'fungus ball'. In neutropenic patients, *Aspergillus* infection typically begins in the lungs and may be followed by fatal disseminated disease. Other sites, including the paranasal sinuses, skin, central nervous system and eye, may become infected; infection at these sites can have a poor prognosis.

Laboratory diagnosis

Sputum culture is of limited value. An isolate from bronchoalveolar lavage is diagnostic (98% specificity) but lacks sensitivity.

Antibody detection may confirm bronchopulmonary aspergillosis and farmer's lung, but immunocompromised patients rarely produce an antibody response. EIA to detect galactomannan in serial samples is useful. NAAT is a useful adjunct to diagnosis.

Treatment and prevention

Bronchopulmonary aspergillosis requires treatment of the airway obstruction with bronchodilators and steroids. Invasive aspergillosis requires treatment with amphotericin B. Itraconazole has activity against *Aspergillus*, and voriconazole improves outcome in pulmonary aspergillosis. Surgery may be beneficial in some cases of pulmonary infection. Patients with farmer's lung should avoid further exposure. Neutropenic patients should be managed in rooms with filtered air and infection aggressively treated.

Other infections

Filamentous fungi may infect severely immunocompromised patients, elderly people, those with poorly controlled diabetes and chronic alcoholics. Sinus infection may spread to the eyes and brain. Pulmonary disease may be complicated by dissemination. *Mucor*, *Rhizopus* and *Absidia* are the main species implicated. Each is difficult to treat and has a poor prognosis.

Dermatophytes

Three species of filamentous fungi are implicated in dermatophytosis: *Epidermophyton*, *Microsporum* and *Trichophyton*. Dermatophytes are also grouped according to their reservoir and host preference: anthrophilic (mainly human pathogens); zoophilic (mainly infect animals); and geophilic (found in soil and able to infect animals or humans). Anthrophilic species spread by close contact (e.g. within families, enclosed communities). Transmission of geophilic species is rare. Close contact with animals may give rise to zoophilic infection (e.g. pet owners, farmers and vets).

Clinical features

Dermatophyte infection may present as red scaly patch-like lesions which spread outwards leaving a pale, healed centre (ringworm). Lesions are itchy but rarely painful, but zoophilic species, produce an intense inflammatory reaction with pustular lesions or an inflamed swelling (kerion). Chronic nail infection produces discoloration and thickening, whereas scalp infection is often associated with hair loss and scarring. Clinical diagnostic labels are based on the site of infection, for example tinea capitis (head and scalp), tinea corporis (trunk lesion), tinea pedis (athlete's foot).

Laboratory diagnosis

Infection of skin and hair by some species may demonstrate a characteristic fluorescence when examined under ultraviolet light (Wood's light).

Skin scrapings, nail clippings and hair samples should be sent dry to the laboratory. When heated in a solution of potassium hydroxide, they clarify and branching hyphae can be seen under the microscope. Dermatophytes grow on Sabouraud's agar at 30°C in 4 weeks.

Identification is based on colonial morphology, microscopic appearance, physiological biochemical testing and sequencing of the 18SrRNA gene.

Treatment

Dermatophyte infections may be treated topically with imidazoles (e.g. clotrimazole). More troublesome infection may require oral treatment (e.g. terbinafine or griseofulvin).

Antifungal compounds
Terbinafine

Terbinafine inhibits squalene epoxidase with resultant accumulation of aberrant and toxic sterols in the cell wall. It is indicated for the oral treatment of superficial dermatophyte infections which have failed to respond to local therapy. Stevens–Johnson syndrome and toxic epidermal necrolysis and hepatic toxicity are reported adverse effects. Treatment should be continued for up to 6 weeks for skin infections and 3 months or longer for nail infections.

Griseofulvin

Griseofulvin is active only against dermatophytes by inhibiting fungal mitosis. Given orally, it is incorporated into the stratum corneum or nail where it inhibits fungal invasion of new skin and nail. Treatment must be continued until uninfected tissue grows. It is now rarely used.

Polyenes

There are two polyene cyclic macrolides in clinical use, nystatin and amphotericin B, which are active against almost all fungi. Polyenes bind ergosterol in the fungal membrane forming a pore, which leads to leakage of the intracellular contents and cell death. Resistance is rare.

Nystatin is used for topical treatment and the prevention of fungal infection in immunocompromised patients. It has no value for the treatment of dermatophyte infections. Amphotericin is given parenterally. The older formulations are relatively toxic, causing fever, chills, thrombophlebitis and hypotension. The drug also causes renal tubular damage that is usually reversible. Lipid formulations are much less toxic and higher doses can be given safely.

Echinocandins

The echinocandins act by inhibiting the synthesis of 1, 3-β-glucan, a homopolysaccharide in the cell wall of many pathogenic fungi. They are active against both *Candida* and *Aspergillus*. New agents of this class (e.g. caspofungin) are entering clinical use, are well tolerated and are a useful alternative for refractory infections.

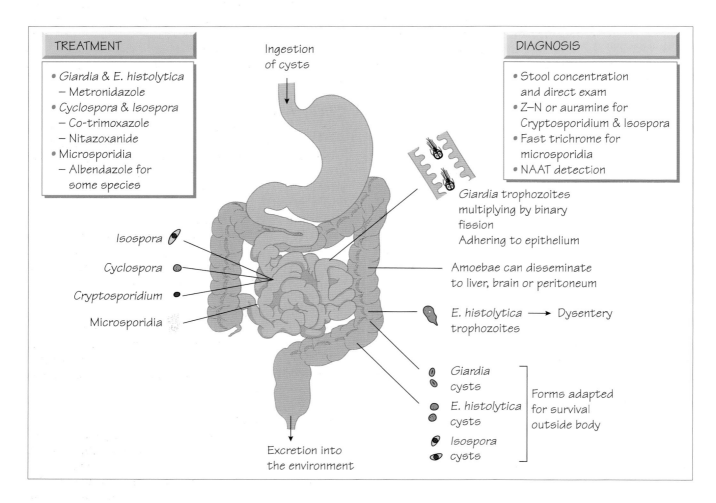

Entamoeba histolytica

Entamoeba histolytica infects the large intestine. Although it is more common in developing countries, cases of amoebiasis are found worldwide. It is transmitted by the faecal–oral route and in food and water. Once ingested, the organisms adhere to the intestinal epithelium and produce proteases and amoeba-pore, an epithelial cytotoxin. It is morphologically identical to *E. dispar* which does not cause disease but is found in the human gut.

Clinical features

The onset is insidious with little systemic upset: the patient is ambulant but has frequent small-volume bloody stools with an offensive odour. Abscesses may develop in the liver or, more rarely, abdomen, lung or brain.

Diagnosis

Sigmoidoscopy reveals rectal ulceration; trophozoites can be demonstrated in ulcer biopsies. Three stool specimens should be sent for cyst identification; rarely, trophozoites are found in fluid stools examined immediately. Antigen detection and NAAT methods enable *E. histolytica* and *E. dispar* to be distinguished. CT and ultrasound may reveal abscesses. Serology is useful in detecting abscesses in the liver or rarely elsewhere, but not intestinal infection.

Treatment

Metronidazole is effective in treating amoebic dysentery but does not eradicate the cyst stage which requires diloxanide furoate or paromomycin. Amoebic abscesses can usually be treated with metronidazole. Surgical drainage is usually unnecessary but is used to prevent abcess rupture.

Prevention and control

Steps to ensure that water is boiled and food adequately cooked will reduce the risk of amoebic infection.

Giardia lamblia

Infection with *Giardia lamblia* is common throughout the world. It occurs where poor sanitation allows water supplies or food to be contaminated with *Giardia* cysts from human, or possibly animal, faeces.

Pathogenesis

Trophozoites multiply in the jejunum by binary fission. They attach themselves strongly to the intestinal wall by a sucking disk. The mechanism for *Giardia* diarrhoea remains unknown but it may be due to direct cytotoxicity, induction of apoptosis or increasing epithelial permeability. Giardial cysts, a form adapted for long-term survival in the environment, are excreted in the faeces.

Clinical features

Infection with *G. lamblia* is characterized by anorexia, crampy abdominal pain, borborygmi and flatus accompanied by bulky offensive fatty stools. Patients may lose weight and there may be an associated lactose intolerance or fat malabsorption. Patients with IgA deficiency may suffer recurrent attacks of infection.

Laboratory diagnosis

Three stools should be examined and concentrated, as the shedding of *Giardia* cysts is intermittent. Aspirated jejunal contents can be examined immediately for the presence of motile trophozoites. EIA and NAAT methods are more sensitive than microscopy.

Treatment

Metronidazole or tinidazole are used. New therapeutic options include albendazole and nitazoxanide. Secondary malabsorption and vitamin deficiency may require investigation and treatment.

Cyclospora cayetanensis

This organism has been recognized as a cause of human diarrhoea. Infection occurs worldwide and outbreaks related to contaminated water supplies and contaminated imported soft fruit and fresh herbs have been reported.

Pathogenesis

Cyclospora are found inside vacuoles within the epithelium of the jejunum. There is inflammation, villous atrophy and crypt hyperplasia leading to malabsorption of B_{12}, folate, fat and D-xylose.

Clinical features

Infection takes the form of watery diarrhoea preceded by a flu-like illness and weight loss. It is self-limiting, but may last for weeks with continuing fatigue, anorexia and weight loss. In HIV-positive individuals, disease is severe, prolonged and relapsing.

Diagnosis and treatment

Diagnosis is by demonstrating oocysts in stools directly or using modified acid-fast stains. NAAT methods are available. Co-trimoxazole is an effective treatment, with nitazoxanide as an alternative.

Cryptosporidium

Cryptosporidium parvum is a zoonotic coccidian parasite that is transmitted by milk, water and direct contact with farm animals. It is naturally resistant to chemical disinfectants, surviving water purification. Person-to-person spread can occur with intimate contact. Infection is common in children and HIV-positive individuals. It may interfere with the glucose-stimulated sodium pump in the small intestine, leading to fluid secretion.

Clinical features

Cryptosporidiosis is usually a self-limiting watery diarrhoea with abdominal cramps. In immunocompromised individuals, diarrhoea is more profuse and prolonged and may cause life-threatening fluid and electrolyte imbalance. Biliary tree, gallbladder and respiratory tract involvement may occur.

Diagnosis and treatment

Cysts are demonstrated in the stool by microscopy using modified acid-fast staining, antigen detection or NAAT. Nitazoxanide may improve clearance of pathogens but management should aim to reverse immuodeficiency.

Isospora belli

A coccidian parasite closely related to *Cryptosporidium*, *Isospora belli* presents with a similar clinical picture, usually following tropical travel. Ziehl–Neelsen's stain of stool identifies the characteristic oval cysts. Treatment is with co-trimoxazole; fluoroquinolones or nitazoxanide are alternatives.

Microsporidia

The microsporidia are small protozoan pathogens of insects, plants and animals. Organisms are intracellular, depending on host cells for a source of energy. They infect neighbouring cells using a long polar tube through which they inject their DNA. *Enterocytozoon bieneusi*, *Encephalitozoon cuniculi*, *Encephalitozoon hellem*, *Septata intestinalis*, *Pleistophora* and *Nosema* have been implicated in human infection.

Pathogenesis

Enterocytozoon bieneusi and *Septata intestinalis* infect epithelial cells of the small bowel, and are associated with diarrhoea. *Encephalitozoon cuniculi* infects macrophages, epithelial cells, vascular endothelial cells and renal tubular cells in the brain and the kidney. It is associated with hepatitis, peritonitis, diarrhoea, seizures and disseminated infection. Before the advent of HIV infection, microsporidia infection was very rare.

Diagnosis and treatment

Microscopy using fast trichrome, calcofluor white and Ziehl–Neelsen stains can be used. Sensitive NAATs are available to demonstrate organisms.

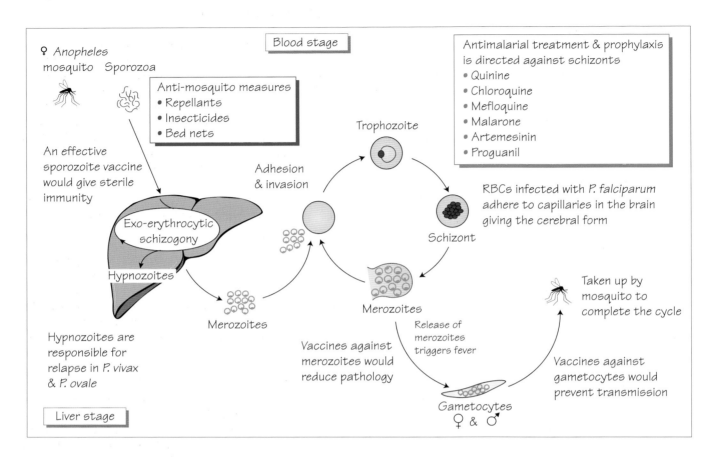

Malaria

Malaria is caused by four species of the genus *Plasmodium*: *P. falciparum*, *P. vivax*, *P. ovale* and *P. malariae*. More than 1.5 billion people live under the threat of malaria; 1 million children under the age of 5 years die each year in Africa alone. In the UK there are more than 2000 reported cases every year and up to 10 deaths. Immigrants returning to their home country are at high risk as they have lost their natural immunity and often omit taking prophylaxis.

Life-cycle

Sporozoa are injected into the circulation by female *Anopheles* mosquito bites. The parasites multiply within hepatocytes. The parasites invade red blood cells (RBCs) and multiply. The parasites provoke the release of cytokines, which are responsible for many of the signs and symptoms of malaria. Infected RBCs develop knob-like projections making them adhere to the capillary wall. This may occur in the brain, causing cerebral malaria.

Some parasites differentiate into the sexual stages, gametocytes, which are taken up by biting female mosquitoes where they develop in the mosquito gut into sporozoites which migrate to the insect salivary glands ready for another bite. *Plasmodium vivax* and *P. ovale* develop dormant stages (hypnozoites), which cause relapses.

Clinical features

Malaria must be considered in any ill patient with a history of travel in an endemic area, particularly if they have fever or flu-like symptoms. Infection by *P. falciparum* can rapidly progress to death, especially in the non-immune traveller; infection by the other species is usually more benign. Holiday travellers have no immunity and regular fevers may not develop. *Plasmodium falciparum* affects every organ and so gives rise to a wide range of complications, such as cerebral malaria, circulatory shock, acute haemolysis and renal failure, hepatitis and pulmonary oedema.

Diagnosis

At least three blood films (both thick and thin) should be obtained at different times, during or directly after a period of fever. Antigen detection dipsticks can also be used for rapid diagnosis and NAAT is useful, especially for detecting drug resistance.

Treatment

Chemotherapy kills the blood stages of the parasites and resistance patterns mean treatment advice must be changed regularly. Combination therapy is the norm; for example, for *Plasmodium falciparum*, quinine, pyrimethamine and sulfadoxine, or quinine and doxycycline. Artemether in combination is also used.

Chloroquine is used for *P. vivax*, *P. ovale* and *P. malariae* infection, and primaquine to eradicate the hypnozoites of *P. vivax* and *P. ovale*.

Prevention and control

Those at risk should sleep under bed nets, cover exposed skin between dusk and dawn when mosquitoes are active and use mosquito repellents. Prophylaxis should be taken following expert up-to-date advice, but remember that patients taking prophylaxis may still develop malaria.

Several vaccines which are in development are directed mainly against the sporozoa. Vaccines against RBC stages and the gametocytes, plus combinations, will probably be required.

Leishmaniasis

Visceral disease is caused by *Leishmania donovani*, *L. infantum* or *L. chagasi*. Cutaneous disease is caused by several species, including *L. major*, *L. tropica* and *L. aethiopica* in the Old World and *L. braziliensis* and *L. mexicana* in the Americas.

Life-cycle

Leishmaniasis is transmitted by sandflies: *Phlebotomus* in the Old World and *Lutzomyia* in the Americas. Sandflies inject the infective promastigotes, which survive ingestion by macrophages where they become amastigotes, multiplying inside cells of the reticuloendothelial system.

Clinical features

In visceral disease, cytokine release by macrophages gives rise to fever and general wasting. Bone marrow is replaced by parasites so the patient becomes anaemic, leucopenic and thrombocytopenic. Reactive hypergammaglobulinaemia makes patients susceptible to secondary bacterial infections: untreated patients will deteriorate and die within 2 years.

Cutaneous forms are characterized by chronic granulomatous lesions at the site of the bite, with satellite lesions. *Leishmania braziliensis* infection causes cutaneous disease and, in some, destruction of structures around the mouth and nose: espundia.

Diagnosis and treatment

The demonstration of parasites in a skin biopsy, bone-marrow sample, blood sample or splenic aspirate by microscopy and culture confirms the clinical diagnosis. NAAT is used for primary diagnosis and speciation. Urinary antigen detection can be used in developing countries. Visceral and cutaneous leishmaniasis can be treated with parenteral liposomal amphotericin B.

Alternatives include antimony compounds, paromomycin and oral miltefosine.

Trypanosomiasis
African trypanosomiasis

African trypanosomiasis is caused by *Trypanosoma brucei gambiense* and *Trypanosoma brucei rhodesiense* and is transmitted by the tsetse fly. Humans are the only host of *T. brucei gambiense*, but antelope or cattle act as the reservoir for *T. brucei rhodesiense*. Parasites in the blood are inhibited by immune responses, but surface antigens change and the organisms multiply again. Generalized lymphadenopathy may be present and the skin may appear oedematous. The patient exhibits a hypergammaglobulinaemia and is susceptible to secondary bacterial infection. When parasites invade the brain they cause a chronic progressive encephalitis: the patient lapses into coma, and death is often the result of secondary bacterial pneumonia.

Diagnosis and treatment

Parasites are demonstrated in blood, CSF or lymph-node aspirate. Serological tests are available. Lumbar puncture should only be performed after circulating parasites have been eliminated with suramin, avoiding the risk of inoculation. The cerebral complications must be treated with melarsoprol (MelB) which itself can cause serious toxicity. Eflornithine can be used in West African disease

South American trypanosomiasis

Trypanosoma cruzi, which causes Chagas' disease, is transmitted by the bite of reduviid bugs. There are three phases of the disease: acute infection characterized by cutaneous oedema, intermittent fever, shock and a significant mortality in children; latent infection; and late manifestations, such as achalasia, megacolon, cardiac dysrhythmias, cardiomyopathy and neuropathy.

Diagnosis

Parasites are demonstrated by microscopy, or culture in artificial medium or laboratory bugs (xenodiagnosis). Serological tests are available.

Treatment

Nifurtimox and benznidazole may be used in the acute phase of infection. Treatment of complications is mainly palliative (e.g. cardiac pacemakers for heart block secondary to cardiomyopathy, surgery for megacolon).

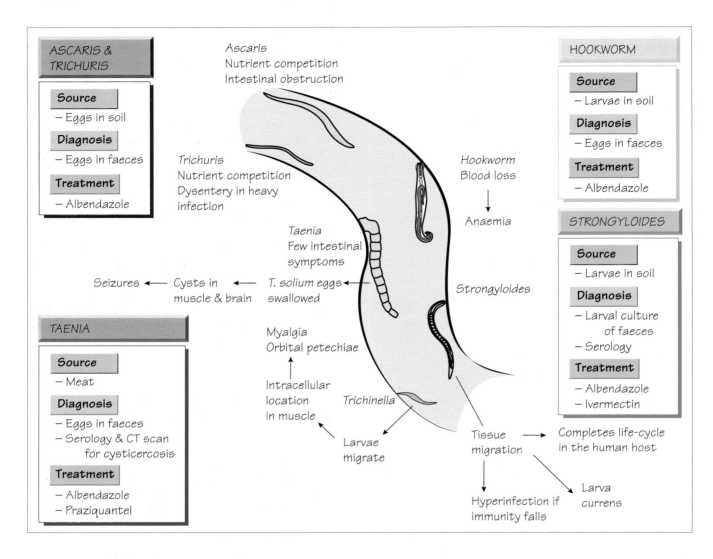

ASCARIS & TRICHURIS

Source
– Eggs in soil

Diagnosis
– Eggs in faeces

Treatment
– Albendazole

Ascaris
Nutrient competition
Intestinal obstruction

Trichuris
Nutrient competition
Dysentery in heavy infection

HOOKWORM

Source
– Larvae in soil

Diagnosis
– Eggs in faeces

Treatment
– Albendazole

Hookworm
Blood loss
↓
Anaemia

STRONGYLOIDES

Source
– Larvae in soil

Diagnosis
– Larval culture of faeces
– Serology

Treatment
– Albendazole
– Ivermectin

Taenia
Few intestinal symptoms

Seizures ← Cysts in muscle & brain ← T. solium eggs swallowed

Strongyloides

TAENIA

Source
– Meat

Diagnosis
– Eggs in faeces
– Serology & CT scan for cysticercosis

Treatment
– Albendazole
– Praziquantel

Myalgia
Orbital petechiae

Intracellular location in muscle

Trichinella

Larvae migrate

Tissue migration → Completes life-cycle in the human host

Hyperinfection if immunity falls

Larva currens

Roundworms and hookworms

Infection by nematodes (roundworms), *Ascaris lumbricoides* and *Trichuris trichiura*, or the hookworms, *Necator americanus* and *Ancylostoma duodenale*, is prevalent in developing countries. Both *Ascaris* and *Trichuris* are acquired by the ingestion of the roundworm eggs, but hookworm larvae invade through intact skin.

Epidemiology

Adult *Ascaris* are found in the intestine; the female worm produces up to 2000 eggs per day. The tough eggs survive in the soil where they mature into infective forms, which may be ingested. After hatching in the small intestine, the organism undergoes a migratory cycle through the liver and lungs, where it may be coughed up and swallowed, eventually to develop into the adult worm resident in the intestine. Transmission occurs in conditions where there is poor sanitation or when food crops are manured with human faeces. In warm, moist climates, the eggs can survive for many years in the soil. Hookworm eggs hatch into an infective larva that is able to burrow through intact skin to cause infection.

Pathogenesis

These parasites cause disease by competing for nutrients; thus, the severity of symptoms is proportional to the number of worms present (parasite load). Hookworm infections are more serious as they take blood, leading to iron-deficiency anaemia that can be severe. Heavily infected children have poor growth and lowered school performance, attributed to micronutrient deficiency, especially with *Trichuris* infection.

Clinical features

Infection is usually asymptomatic, but heavy *Ascaris* infection may lead to intestinal obstruction, and heavy *Trichuris* infection to a dysentery-like syndrome.

Diagnosis

The diagnosis is made by examining up to three stool samples for the presence of the characteristic eggs.

Treatment

Intestinal nematodes can be treated with imidazoles such as albendazole or mebendazole. Improved sanitation is required to control the spread of infection.

Threadworms

Humans are the only host of *Enterobius vermicularis*. Living in the large intestine, the females migrate to the anus where they lay their eggs on the perianal skin. Symptoms are more common in children but often the whole family is infected. Symptoms are few; thread-like worms may be found in the faeces or patients may complain of perianal itching, often worse at night. Scratching allows contamination of the fingers with larvae containing eggs which, when placed in the mouth, initiate a new cycle of infection. Occasionally, *Enterobius* can be found in the appendix.

Diagnosis is made by sending an adhesive tape swab to the laboratory where D-shaped eggs are seen. Treatment is with mebendazole or piperazine. It is often necessary to treat the whole family, repeating treatment after 2 weeks, and hygiene should also be addressed.

Strongyloides stercoralis

Strongyloides stercoralis larvae are passed in the stools, where they either undergo a free-living cycle in the soil or differentiate into infective larvae that invade another host via intact skin. Inside the human host, they can initiate another development cycle. The consequence of this is that infection with *Strongyloides* can be prolonged. Resistance to *Strongyloides* depends on efficient cell-mediated immunity. Individuals who are infected with human T-cell leukaemia virus 1 or are taking steroids are especially susceptible to hyperinfection.

Clinical features

Migrating larvae leave a red itchy track, which fades after about 48 h. If the patient is given immunosuppressive therapy, uncontrolled multiplication of the *Strongyloides* may occur, characterized by fever, shock and the signs of septicaemia and meningitis.

Diagnosis

Stool culture from multiple specimens may reveal infective larvae. Alternatively, samples of jejunal fluid are examined for the presence of larvae. A sensitive EIA technique for serum is available.

Treatment

Ivermectin is the optimal drug used for treatment, with the imidazoles (e.g. albendazole) as alternatives. Relapse occurs in up to 20% of patients. The hyperinfection syndrome is often accompanied by Gram-negative septicaemia, which requires urgent treatment.

Prevention

The risk of infection can be reduced by wearing appropriate footwear to prevent larvae penetrating the skin.

Tapeworms

Taenia spp.

Two *Taenia* spp. infect humans: the pork worm, *Taenia solium*, and the beef worm, *Taenia saginata*. Infection is acquired by eating the meat of intermediate hosts that contains the tissue stages of the parasite.

Pathogenesis and clinical features

Tapeworms compete for nutrients and infections are usually asymptomatic.

Taenia solium can use humans as an intermediate as well as the definitive host. When an individual ingests *T. solium* eggs, they hatch and disseminate, forming multiple cyst-like lesions in the muscles, skin and brain. These 'measly' lesions, similar in appearance to infected pork meat, are known as cysticercosis. Inflammatory responses to parasitic antigens leaking from cysts in the brain may lead to epileptic seizures.

Diagnosis

This is made by finding characteristic eggs in the patient's stool. Cysticercosis is diagnosed by a specific EIA and confirmed by demonstrating the presence of multiple tissue cysts by X-ray, CT or MRI.

Treatment and prevention

Treatment is with praziquantel. Specialist advice should be sought for the management of *Taenia* infections in the central nervous system as severe inflammatory reactions can occur.

Diphyllobothrium latum

Humans are the definitive host of this tapeworm, acquiring infection from undercooked freshwater fish. The parasite competes for nutrients and causes deficiency of vitamin B_{12}. The diagnosis is made by detecting characteristic eggs in faeces. Treatment is with praziquantel.

Hymenolepis nana

Humans are the only host of this small tapeworm. Infection is usually asymptomatic and diagnosis is made by detecting the characteristic eggs in the faeces. Treatment is with praziquantel.

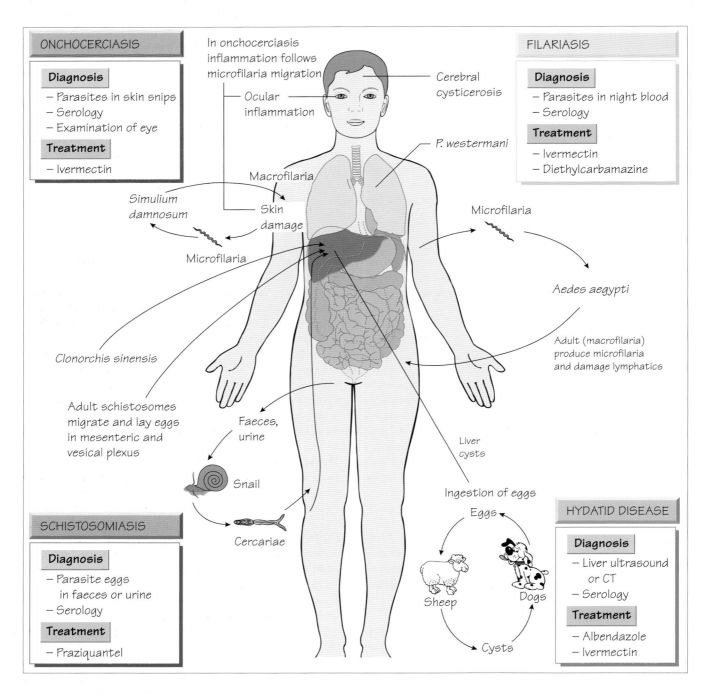

ONCHOCERCIASIS

Diagnosis
– Parasites in skin snips
– Serology
– Examination of eye

Treatment
– Ivermectin

In onchocerciasis inflammation follows microfilaria migration

Ocular inflammation

Cerebral cysticerosis

FILARIASIS

Diagnosis
– Parasites in night blood
– Serology

Treatment
– Ivermectin
– Diethylcarbamazine

P. westermani

Macrofilaria

Simulium damnosum

Skin damage

Microfilaria

Microfilaria

Aedes aegypti

Clonorchis sinensis

Adult (macrofilaria) produce microfilaria and damage lymphatics

Adult schistosomes migrate and lay eggs in mesenteric and vesical plexus

Faeces, urine

Snail

Liver cysts

Ingestion of eggs

Eggs

HYDATID DISEASE

Diagnosis
– Liver ultrasound or CT
– Serology

Treatment
– Albendazole
– Ivermectin

SCHISTOSOMIASIS

Diagnosis
– Parasite eggs in faeces or urine
– Serology

Treatment
– Praziquantel

Cercariae

Sheep

Dogs

Cysts

Schistosomiasis

Three species infect humans: *Schistosoma mansoni* (Africa and South America); *S. japonicum* (Far East); and *S. haematobium* (Africa). Eggs are excreted in the faeces and urine of infected humans. In areas of poor sanitation, the eggs hatch, releasing a miracidium which invades a snail. After development in the snail, the schistosome cercariae emerge into the environment. They actively penetrate intact skin and develop into male and female adult worms that migrate to the superior, inferior mesenteric or vesical plexus, depending on species, where they lay eggs.

Pathogenesis and clinical features

Initial infection is characterized by fever, hepatosplenomegaly, skin rash and arthralgia. Later bloody diarrhoea or haematuria occurs, associated with egg expulsion. Years later symptoms and signs are caused by the fibrotic reaction to eggs, for example in the liver (hepatic fibrosis and portal hypertension), the lungs (lung fibrosis) and the bladder (in the case of *S. haematobium*). Space-occupying lesions in the brain and spinal cord may lead to seizures.

Diagnosis

Eggs can be demonstrated in stool, urine, rectal snips or other tissue biopsy. An EIA detecting antischistosomal antibody is useful especially in travellers, and antigen detection methods are available as research tests.

Prevention and control

Infection is prevented by appropriate clothing when working in the fields and avoiding contaminated water. Control programmes targeting snails, or mass treatment, can control the disease if sufficient resources are available, as has been demonstrated in China and Japan.

Filariasis

Lymphatic filariasis is caused by *Brugia malayi* and *Wuchereria bancrofti* and transmitted by the mosquito *Aedes aegypti* throughout the tropics. Onchocerciasis is caused by *Onchocerca volvulus*, and transmitted by the blackfly *Simulium damnosum* in West Africa and South and Central America. Loa (loiasis) is caused by *Loa loa* and transmitted by *Chrysops* flies in West Africa.

Clinical features

Inflammation arises from the response to endobacteria (*Wolbachia* spp., related to *Rickettsia*) that infect the filariae.

Lymphatic filariasis is characterized by acute attacks of fever and lymphoedema, which may be complicated by secondary bacterial infection. After repeated attacks lymphatic vessels are permanently damaged, leading to lymphoedema in the leg, arm or scrotum.

Onchocerca adults are located in nodules and microfilariae migrate in the skin resulting in pruritus and dry thickened skin. Inflammation in the eye causes blindness.

Loiasis is less damaging and diagnosis is based on fleeting subcutaneous swellings, known as Calabar swellings. Infection may be associated with fever and abnormalities of renal function.

Diagnosis

Lymphatic filariasis is diagnosed by identifying microfilariae in the peripheral blood at midnight. Blood is filtered and the filter stained and examined microscopically. *Loa loa* is detected in daytime blood films; other filariae appear at night.

In onchocerciasis 'pinch biopsies' should be taken from any affected area, plus the shoulder-blade, the buttocks and thighs. On microscopic examination, microfilariae can be seen to emerge from the skin. If biopsy is negative, a 50-mg dose of diethylcarbamazine will induce increased itch (the Mazzotti reaction). Pinch biopsies taken at this time are more likely to be positive. EIA is also used for diagnosis.

Treatment

Lymphatic filariasis is treated with diethylcarbamazine or ivermectin together with albendazole. Ivermectin is the treatment of choice for onchocerciasis, and the addition of tetracycline enhances the effect on microfilarial load by sterilizing the adult worms. Treatment of filarial infection may stimulate an acute inflammatory reaction.

Prevention and control

Onchocerciasis is an important public health problem in West Africa. An international control programme is under way, using mass treatment of whole populations with ivermectin and doxycline. Lymphatic filariasis is prevented by mosquito control measures. An international control programme for Guinea worm has brought about its effective eradication.

Hydatid disease
See Chapter 52.

Clonorchis sinensis (*Opisthorchis sinensis*)

Infection is acquired by eating undercooked fish containing metacercariae. The adults live in the bile ducts, and eggs are passed in the faeces. Infection is found mainly in the Far East, but can occur wherever infected fish is imported. Light infections are usually asymptomatic, but heavier infections result in cholangitis and pancreatitis, and biliary obstruction and cirrhosis may develop. Cholangiocarcinoma is a late complication. The diagnosis is made by identifying the characteristic eggs in faeces. Patients may be treated with praziquantel. Infection is prevented by adequate cooking of potentially infected fish.

Fasciola hepatica

Humans are accidental hosts of this sheep and cattle parasite. The infective stage is found on freshwater plants such as watercress which, if eaten without cooking, can result in infection. The larvae hatch in the intestine; after maturation and migration, the adults are located in the liver. Patients present with fever and right upper quadrant pain, but these symptoms abate. Low-grade biliary symptoms and liver fibrosis may denote continuing infection. Treatment is with praziquantel.

Paragonimus spp.

Parasites of this genus infect different organs: *P. westermani*, the lungs; and *P. mexicanus*, the brain. This rare infection follows the ingestion of undercooked crustaceans. Acute non-specific symptoms such as fever, abdominal pain and urticaria are followed by specific symptoms and signs such as chest pain, dyspnoea and haemoptysis or central nervous system signs. Diagnosis is by identifying characteristic eggs in sputum, imaging, and serology or tissue biopsy. Lung fluke is treated with praziquantel; cerebral disease with a combined surgical and medical approach.

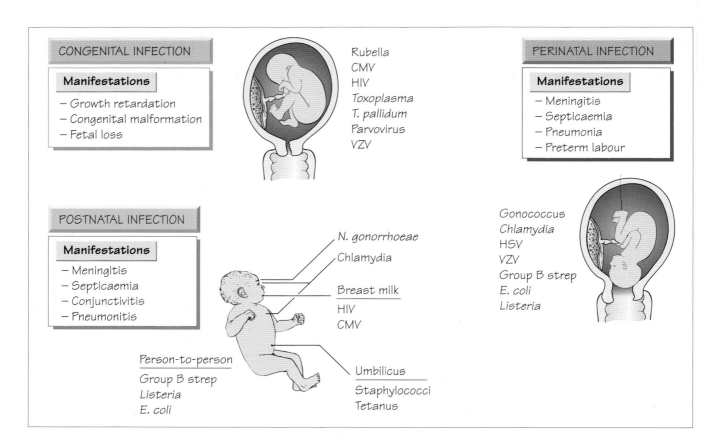

CONGENITAL INFECTION

Manifestations

- Growth retardation
- Congenital malformation
- Fetal loss

Rubella
CMV
HIV
Toxoplasma
T. pallidum
Parvovirus
VZV

PERINATAL INFECTION

Manifestations

- Meningitis
- Septicaemia
- Pneumonia
- Preterm labour

POSTNATAL INFECTION

Manifestations

- Meningitis
- Septicaemia
- Conjunctivitis
- Pneumonitis

N. gonorrhoeae
Chlamydia

Breast milk
HIV
CMV

Gonococcus
Chlamydia
HSV
VZV
Group B strep
E. coli
Listeria

Person-to-person
Group B strep
Listeria
E. coli

Umbilicus
Staphylococci
Tetanus

Infection may be acquired across the placenta (intrauterine infection) or contracted during the process of birth or by direct contact with maternal body fluids. Prolonged rupture of the membranes predisposes to fetal infection. Infection can also be transmitted to the neonate after birth from the mother or other contacts.

Congenital rubella

Jaundice associated with hepatitis is often the first sign of congenital rubella. Haemolysis and thrombocytopenic purpura are also common, as is a low-grade meningoencephalitis. Some babies have evidence of metaphyseal dysplasia. Infected infants have low birth weight and fail to attain their expected developmental milestones. There is a high mortality in severely affected infants. Patent ductus arteriosus, cataracts, deafness and retinal pigment dysplasia may be present.

Rubella IgM is positive and persists until the third month of life. Infection during the first trimester is more than 60% and some parents will opt for termination of pregnancy. Later the risk is much lower (2% after 20 weeks) and the balance between the chance of fetal damage and the desirability of termination should be considered carefully.

Cytomegalovirus

Infection occurs in less than 1% of births, of which 1% are severely affected. The risk of infection is highest during the first trimester. It presents with prematurity, low birth weight, hepatomegaly, splenomegaly, thrombocytopenia and prolonged jaundice, cerebral irritability, fits or abnormal muscle tone or movement.

Microcephaly and sensorineural deafness are the most common problems. Other problems include cerebral calcification, hemiplegia, psychomotor retardation, choroidoretinitis and myopathy. Diagnosis depends on demonstrating IgM antibodies or cytomegalovirus excretion during the first 20 days of life.

Congenital and intrapartum herpes simplex infections

Primary herpes simplex infections may be accompanied by viraemia when transplacental infection can occur. Infants born with congenital infection tend to have severe disease, with pneumonitis, meningoencephalitis, hepatosplenomegaly and cytopenias. Only a few will demonstrate herpetic skin or mucosal lesions. Treatment with aciclovir reduces mortality from 80–90% to 10–15% and should not wait for laboratory confirmation.

Primary infection may be contracted at birth from maternal genital herpes. Skin, conjunctival, oral or genital lesions develop within a few days, with dissemination in 50% of cases. Treatment is with intravenous aciclovir.

Varicella

Varicella embryopathy follows maternal infection during the first or second trimester of pregnancy; it is transmitted in less than 3% of infected pregnancies. Cicatricial contracture of a limb with hypoplasia, microcephaly or microphthalmia may occur. Non-immune women exposed to chickenpox should be offered postexposure prophylaxis with zoster immune globulin (ZIG) within 10 days of exposure.

Neonatal varicella occurs when the mother develops chickenpox within 1 week of delivery. As neonatal mortality is up to 40%, the neonate should be given ZIG within 48 h of birth if possible and treated with acyclovir if infection develops. Normal immunoglobulin given to the mother will not protect the infant. A vaccine is entering clinical use in some countries.

Listeriosis

Transplacental transmission of *Listeria monocytogenes* occurs during a maternal infection that is often inapparent. Infection in early pregnancy often results in fetal death; later infection is associated with premature labour. Severe bacteraemia, associated with hepatosplenomegaly, meningoencephalitis, thrombocytopenia and pneumonitis, usually complicates neonatal infection. Intrapartum exposure may lead to neonatal infection during the first 2 weeks of life, usually with meningitis and bacteraemia. Blood, CSF, placental tissue and lochia should be cultured. Infected mothers and infants may be a source of infections in the postnatal ward and should be isolated. Ampicillin with or without the addition of gentamicin (for 2–6 weeks) is the treatment of choice.

Syphilis

Congenital infection is now rare as a result of antenatal screening. Affected babies are feverish with features similar to secondary syphilis: rash, condylomata and mucosal fissures. Osteochondritis may cause pain. Persistent rhinitis ('snuffles') is common.

Diagnosis is confirmed by dark-ground microscopy of mucosal or skin lesions. Specific IgM or antibodies persisting after 6 months indicate infection. Late manifestations appear between 12 and 20 years: deafness, optic atrophy or paretic neurosyphilis. Other features include bossing of the frontal bones, chronic tibial periostitis, notching of the incisors, 'mulberry' deformity of the first permanent molar and a high arched palate. The treatment of choice is benzylpenicillin.

Toxoplasmosis

The incidence of toxoplasmosis varies internationally; it is uncommon in the UK, but common in France. Transplacental infection occurs in a third of affected pregnancies. Infection in the first and second trimester is more likely to cause significant fetal disease: the fetus may be stillborn, die soon after birth, or have cerebral calcification, cerebral palsy or epilepsy. Chorioretinitis may not be evident until after birth and may be the only feature. Maternal toxoplasmosis is confirmed by specific IgM antibodies or by seroconversion. IgM antibodies may also be demonstrated in affected neonates. Treatment with spiramycin may reduce the risk of transplacental infection but does not affect the outcome of fetal disease.

Perinatal infections
Bacteraemia and pneumonia

In the first few days of life there are few specific clinical features of bacteraemia. The neutrophil count may rise, though this is not always reliable. A fall in platelets, bradycardia and rise in C-reactive protein (CRP) may also occur. Meningitis also presents non-specifically. Blood, urine and CSF culture should be performed but treatment should not wait for laboratory confirmation. Therapy should be targeted at *Escherichia coli* and group B streptococci (e.g. benzylpenicillin and gentamicin, or cefotaxime). In some countries screening for group B streptococci is performed during pregnancy and intrapartum penicillin prophylaxis prescribed; in others screening is targeted to high-risk pregnancies.

Gonococcal ophthalmia neonatorum

Neisseria gonorrhoeae infection may be contracted during delivery, causing ophthalmia neonatorum, a purulent conjunctivitis. It is diagnosed by direct Gram stain and culture. Systemic penicillin will treat the infection. Cefotaxime is used if there is antimicrobial resistance.

Chlamydia

Chlamydial ophthalmia neonatorum is a severe conjunctivitis appearing within 4 days of birth. It is often followed at 6 weeks of age by pneumonitis characterized by tachypnoea and cough. Conjunctivitis is treated with topical tetracycline. Erythromycin is the treatment of choice for chlamydial pneumonitis.

Bullous impetigo (Lyell's syndrome)

Infection with *Staphylococcus aureus* expressing exfoliative toxins results in superficial blisters or bullae which break, leaving extensive raw areas—the 'scalded skin syndrome'.

Treatment with flucloxacillin should be prescribed. The infant and mother should be isolated.

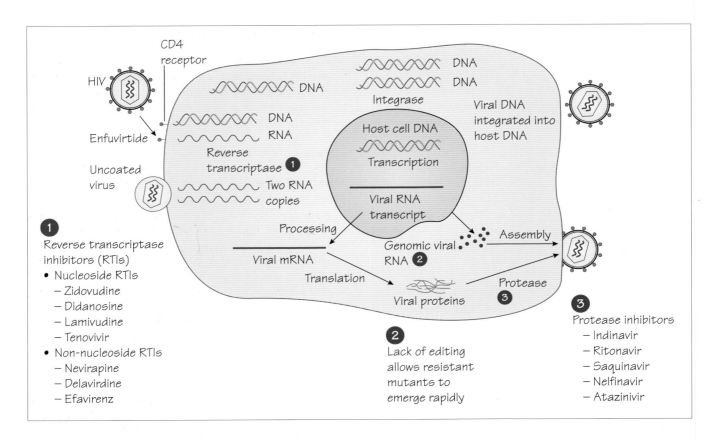

Human immunodeficiency virus

HIV is a spherical, enveloped RNA virus. It is a retrovirus, using reverse transcriptase to produce a DNA copy from viral RNA that is incorporated into the host nucleus to become the template for further viral RNA. Three genes are required for viral replication: *gag*, *pol* and *env*. HIV is classified as a lentivirus. There are two which are pathogenic for humans: HIV-1, which is most common; and HIV-2, which is found mainly in West Africa and appears to be less virulent.

Epidemiology

HIV infection has spread worldwide, transmitted by the parenteral and sexual routes. Infection is most common in patients at high risk of sexually transmitted disease, especially where genital ulceration is common. In developed countries, the main risk groups are men who have sex with men and intravenous drug users. Heterosexual transmission is less common but does occur. In developing countries, HIV spreads mainly by heterosexual transmission and through unscreened transfusions or contaminated medical equipment. Infection can be transmitted from mother to fetus.

Pathogenesis

The virus principally infects cells with a CD4 receptor (e.g. T cells and macrophages). Viral replication results in progressive T-cell depletion and diminished cell-mediated immunity. Different virus strains display different affinities for cells expressing different chemokine receptors. Lacking T-cell help, B-cell function is also reduced. HIV causes damage to neural cells and stimulates cytokine release that may also cause neurological damage. Many of the clinical signs of HIV infection are caused by secondary infections which occur when the CD4 count falls.

Clinical features

A few weeks after infection, a mononucleosis-like syndrome may develop with rash, fever and lymphadenopathy. A latent period follows that may last as long as 10–15 years. When T-cell function is sufficiently compromised ($< 0.2 \times 10^9$/L), secondary infections and malignancies develop as a result of profound immunosuppression, a condition known as acquired immune deficiency syndrome (AIDS), although this is reversed by highly active anti-retroviral therapy (HAART).

Bacteria: *Mycobacterium tuberculosis*, *M. avium-intracellulare* (see Chapter 17), *Salmonella*, *Streptococcus pneumoniae*.

Protozoa: *Toxoplasma gondii*, *Cryptosporidium parvum*, *Isospora belli*, microsporidia.

Fungi: *Candida* spp., *Cryptococcus neoformans*, *Pneumocystis carinii*.

Viruses: varicella zoster, human papovavirus.

Malignancy: Kaposi sarcoma (HHV-8), non-Hodgkin's lymphoma.

Children with HIV are especially vulnerable to childhood virus infections, e.g. measles, and recurrent bacterial infections, e.g. pneumonia.

Diagnosis

Diagnosis is by detection of HIV-specific antibody using two different immunoassay methods, such as EIA, competitive EIA, particle agglutination or Western blotting. Individual testing must always be preceded by counselling. In addition, as sero-conversion may take up to 3 months, an initial negative result should be repeated.

HIV RNA can be detected in clinical samples by NAAT (RT-PCR) in serum. Treatment is monitored by quantitative assays (viral load). HIV may be grown in lymphocytes but this is not used in diagnosis.

Treatment

Agents available to treat HIV infection include nucleoside reverse transcriptase inhibitors (NRTIs), e.g. zidovudine; non-nucleoside reverse transcriptase inhibitors (NNRTIs), e.g. nevirapine; and protease inhibitors, e.g. indinavir. At present, three main principles govern treatment: to minimize viral replication; to prevent emergence of resistant virus; and to reconstitute the immune response. Treatment should start for symptomatic patients or those with opportunistic infections and when CD4 count is $< 0.2 \times 10^9/L$. Between 0.2 and $0.35 \times 10^9/L$ treatment should be considered, bearing in mind the risk of side effects and the likelihood of patient adherence.

There are many possible regimens but initial regimens often contain an NNRTI, and two NRTIs or protease inhibitor and NRTI. Care must be taken if patients are infected with resistant virus as tailored regimens are required. Because RNA viruses lack efficient genetic proof-reading mechanisms, mutations arise rapidly and patients develop drug resistance quickly.

As the immune system starts to recover with treatment, symptoms from opportunistic infection can worsen due to the effects of the enhanced immune reaction.

Prevention

HIV transmission is prevented by avoiding high-risk partners and unprotected intercourse (e.g. by using barrier contraception). Blood products must be screened and potentially HIV-infected material discarded. Health education and free needle exchange programmes may reduce the risk of transmission between intravenous drug users. Antigenic diversity has prevented a vaccine from being developed. Anti-retroviral prophylaxis is required for needle-stick injuries. Transmission from mother to child can be reduced by administration of HAART, elective caesarian section and avoidance of breast feeding. If HAART is not available (in developing countries) a short course of anti-retroveral therapy can reduce the risk of tranmission.

Pneumocystis jiroveci

Pneumocystis jiroveci is a fungus causing infection only in patients who have severe T-cell dysfunction through HIV, malnutrition, prematurity, primary immune deficiency diseases and immunosuppressive drugs. Prior to the HIV epidemic, the infection was rare. Transmitted by the respiratory route, *P. jiroveci* adheres strongly to pneumocytes.

Clinical presentation

Patients typically present with dyspnoea, which develops insidiously over days or weeks, and an unproductive dry cough. Pleuritic chest pain is uncommon. Although patients are febrile, clinical examination is usually normal, although fine basal crackles may be heard. Initially the chest X-ray may appear to be normal but reticular shadowing may develop until there is diffuse air space consolidation. In a small proportion of patients there are atypical features.

Diagnosis

Specimens, obtained by bronchoalveolar lavage or by the use of nebulized hypertonic saline, are examined by specific immunofluorescence, methenamine silver staining or NAAT.

Treatment

Treatment is with oral co-trimoxazole in high dosage, or intravenous pentamidine. Alternatives include trimethoprim–dapsone, pyrimethamine–clindamycin and atovaquone.

Other fungal infections

HIV patients may develop severe mucocutaneous candidiasis with oral ulceration and oesophageal infection. This causes dysphagia with subsequent weight loss. Acute infection is treated with oral agents (e.g. fluconazole), but as long-term suppressive treatment is required, resistance to these agents may develop. Cryptococcal meningitis is a common and recurrent problem.

Toxoplasma gondii

Toxoplasma infection persists inside the host cells for very long periods. Falling immunity allows the reactivation of previous dormant infection. A space-occupying lesion may develop in the brain that may be accompanied by encephalitis.

Toxoplasma encephalitis presents with fever and headaches. Convulsions, coma and focal neurological signs may follow. CT scan may demonstrate multiple diagnostic focal lesions with ring enhancement. Brain biopsy may yield material for tissue culture or PCR. *Toxoplasma* encephalitis is treated with pyrimethamine–sulfadiazine. Long-term suppressive treatment is required after recovery.

45 Pyrexia of unknown origin and septicaemia

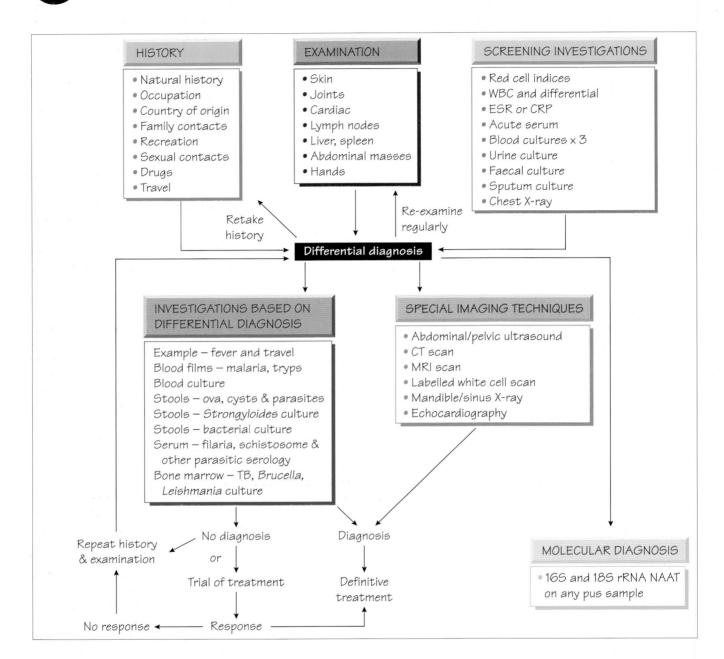

Pyrexia of unknown origin

Definition

A fever, intermittent or persistent, of greater than 38.2°C for more than 2 weeks for which there is no obvious cause is termed pyrexia of unknown origin (PUO).

Aetiology

Infection accounts for 45–55% of cases of PUO. Causes include endocarditis, tuberculosis, osteomyelitis and hidden abscess (e.g. dental or intra-abdominal). Malignancy accounts for 12–20% of cases, commonly lymphomas, or renal cell or lung carcinoma. Connective tissue disorders such as rheumatoid arthritis, systemic lupus erythematosus or polyarteritis nodosum are responsible for 10–15% of cases. Hypersensitivity to drugs, pulmonary emboli, granulomatous diseases (e.g. sarcoidosis), rare metabolic conditions (e.g. porphyria) and a factitious cause (i.e. fever induced deliberately by the patient) are less common causes that should be considered. The longer the history of fever, the more likely it is that it arises from a *non*-infectious cause.

Investigation

When investigating a possible infective cause, a detailed history of the presenting complaint, including the occupational, social and sexual history of the patient, is important. For example, wind-surfers are at risk of leptospirosis, veterinarians and farmers zoonotic infections, such as brucellosis. Recent travel may suggest exposure to unusual tropical infections, some of which have a long incubation period. Sexual history may indicate possible HIV or related risks. The medical history should include a detailed list of medications, whether prescribed by a doctor or not. Many drugs and health products can cause fever. Patients may be reluctant to tell their doctor about drugs purchased from an alternative medicine practitioner. Patients may overlook vital information at the first interview; further opportunities for discussion are often necessary.

The patient must be examined carefully for localized bone or joint pain, subtle rashes, lymphadenopathy, abdominal masses, cardiac murmurs and mild meningism. A complete physical examination should be repeated regularly to detect changes, such as a new soft systolic murmur or increasing abdominal tenderness. A spleen or liver which was impalpable on first examination may have become so because of the development of the condition.

After the initial history and examination is taken the investigative process can be divided into three phases.
1 Screening investigations which are performed on all patients (see figure).
2 Investigations which depend on the results of history and physical examination.
3 Further screening tests and special imaging techniques: ultrasound, CT, MRI, echocardiography and dental X-ray.

The results of the preliminary history and examination are taken together with the results of the primary investigations to plan the tests that will be performed in the second round. These investigations are chosen based on syndrome groups, e.g. fever, eosinophilia and tropical travel. If a diagnosis is not made, further imaging techniques may reveal an occult abdominal abscess or osteomyelitis.

Management

A diagnosis should be made before any antimicrobial therapy is commenced. In some infections, notably tuberculosis, which are suspected but not proved, a trial of therapy may be considered. If this produces clinical improvement, a full course of chemotherapy may be initiated.

Septicaemia
Aetiology

Bacteraemia may arise from normal flora, which have become invasive in conditions such as dental abscess, cholecystitis, appendicitis or diverticulitis. Septicaemia following surgery may be caused by a wide range of organisms, including contamination by skin flora. This problem is particularly important in surgery involving prosthetic devices (orthopaedic, cardiovascular, neurosurgical). The urinary tract is a very common source of Gram-negative infection (see Chapter 49). *Streptococcus pneumoniae* bacteraemia (see Chapter 48) may follow pneumonia; *Streptococcus pyogenes* or *Staphylococcus aureus* bacteraemia may complicate skin infections. Septicaemia caused by *Neisseria meningitidis* or *Streptococcus pneumoniae* may be accompanied by meningitis.

Clinical features

Although the septicaemic patient is usually severely ill with fever and shock, sometimes aggravated by depressed consciousness, septicaemia may be asymptomatic. Fever may be absent in children and elderly people, shock may not have yet developed, and they may present as confused, drowsy or generally unwell. Clinically, it is impossible to distinguish Gram-positive from Gram-negative shock. Some organisms may have characteristic associated clinical signs (e.g. the purpura of *N. meningitidis*).

Diagnosis and treatment

With the exception of suspected *N. meningitidis* infection, at least two specimens of peripheral blood should be taken for culture before therapy is commenced. Other investigations are directed to finding the source of the sepsis, such as urine culture, CSF, abdominal ultrasound, sputum culture and skin swabs. Chest and abdominal X-ray should be performed.

Empirical therapy based on the source of infection and likely infecting organism should be started promptly. Parenteral antibiotics, covering the likely pathogens, should be prescribed.

Puerperal fever

This is a severe, usually bacteraemic, infection caused by entry of pathogens through the placental bed or the cervix within 7 days of delivery. The organisms involved are either primary pathogens introduced by medical manipulation (e.g. *S. pyogenes*) or elements of the normal flora such as *Bacteriodes* and coliforms associated with retained products. Fever, back pain, offensive lochia or shock may be present. Infection may be complicated by disseminated intravascular coagulation. Fever in the early puerperium should be investigated with blood and urine culture and endocervical swabs. Empirical treatment with parenteral antibiotics such as a third-generation cephalosporin and metronidazole or a combination such as piperacillin plus tazobactam should begin without delay. Retained products of conception should be removed. Intensive-care support may be required.

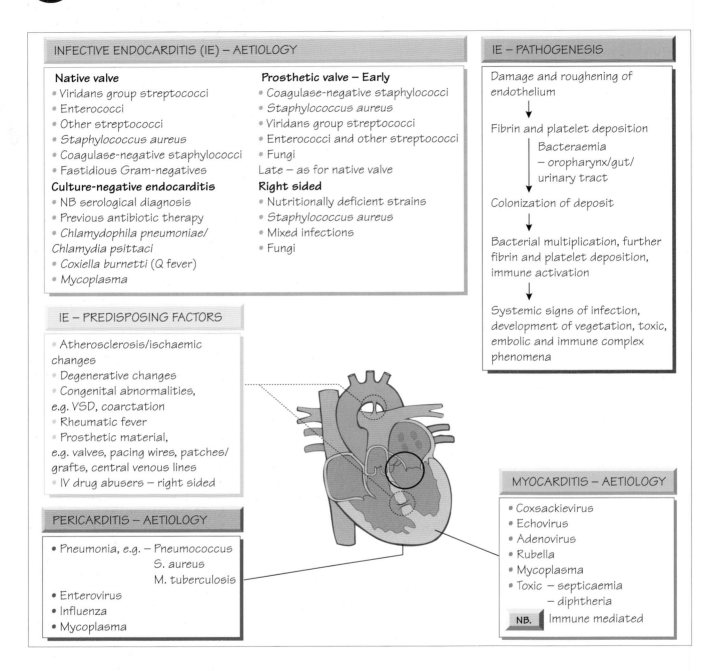

INFECTIVE ENDOCARDITIS (IE) – AETIOLOGY

Native valve
- Viridans group streptococci
- Enterococci
- Other streptococci
- Staphylococcus aureus
- Coagulase-negative staphylococci
- Fastidious Gram-negatives

Culture-negative endocarditis
- NB serological diagnosis
- Previous antibiotic therapy
- Chlamydophila pneumoniae/Chlamydia psittaci
- Coxiella burnetti (Q fever)
- Mycoplasma

Prosthetic valve – Early
- Coagulase-negative staphylococci
- Staphylococcus aureus
- Viridans group streptococci
- Enterococci and other streptococci
- Fungi
Late – as for native valve

Right sided
- Nutritionally deficient strains
- Staphylococcus aureus
- Mixed infections
- Fungi

IE – PATHOGENESIS

Damage and roughening of endothelium
↓
Fibrin and platelet deposition
Bacteraemia
– oropharynx/gut/urinary tract
↓
Colonization of deposit
↓
Bacterial multiplication, further fibrin and platelet deposition, immune activation
↓
Systemic signs of infection, development of vegetation, toxic, embolic and immune complex phenomena

IE – PREDISPOSING FACTORS

- Atherosclerosis/ischaemic changes
- Degenerative changes
- Congenital abnormalities, e.g. VSD, coarctation
- Rheumatic fever
- Prosthetic material, e.g. valves, pacing wires, patches/grafts, central venous lines
- IV drug abusers – right sided

PERICARDITIS – AETIOLOGY

- Pneumonia, e.g. – Pneumococcus
  S. aureus
  M. tuberculosis
- Enterovirus
- Influenza
- Mycoplasma

MYOCARDITIS – AETIOLOGY

- Coxsackievirus
- Echovirus
- Adenovirus
- Rubella
- Mycoplasma
- Toxic – septicaemia
  – diphtheria
- NB. Immune mediated

Endocarditis

Heart valves may be infected during transient bacteraemia. Congenitally abnormal or damaged valves are at greatest risk. Bacteria may originate from the mouth, urinary tract, intravenous drug misuse or colonized intravascular lines.

Clinical features

Patients present with malaise, fever and variable heart murmurs. Arthralgia is sometimes present. The classical stigmata, e.g. splinter haemorrhages, Osler's nodes, microhaematuria, retinal infarcts, finger clubbing, *café-au-lait* skin, Janeway's lesions, are only seen when infection has been present for some time. In later stages, septic emboli may cause a stroke. With more virulent organisms such as *Staphylococcus aureus*, infection progresses rapidly and signs of acute sepsis may predominate.

Diagnosis

Most use a variation of the 'Duke' criteria for diagnosis. Major criteria include positive blood culture for typical organisms (e.g. viridans streptococci), persistently positive blood cultures with any organism, evidence of endocardial involvement demonstrated by echocardiogram and new valvular regurgitation. Minor criteria include predisposition, fever > 38°C, immunological signs (e.g. septic pulmonary infarcts), and echocardiographic or microbiological evidence not meeting major criterion. A diagnosis is made if there are two major criteria present or one major and three minor.

Complications

Local progression may lead to abscess formation in the aortic root. Destruction of the valve results in rapid cardiac decompensation and severe cardiac failure. Cerebral or limb infarction may follow septic embolus. Nephritis is secondary to immune complex deposition and can progress rapidly if sepsis is uncontrolled or renal-toxic antibiotics are given without care (e.g. aminoglycosides).

Investigation

Echocardiography, either transthoracic or trans-oesophageal (more sensitive), will demonstrate vegetations on the valves; a plain chest X-ray may show evidence of cardiac failure. At least three sets of blood cultures should be taken, an hour apart, while fever is present. Antibiotic therapy should await the results of blood culture if possible. Serum should be tested for antibodies to *Coxiella* and *Chlamydia psittaci*.

Management

Ideally, antibiotics should not be commenced until the identity and sensitivities of the infecting organism are known; the prognosis of empirically treated, culture-negative endocarditis is poorer than when the infecting organism is identified. Careful microbiological monitoring of the markers of inflammation (e.g. CRP) is associated with an improved outcome.

Therapy should be planned, based on sensitivity testing and following determination of minimum inhibitory concentration (MIC) and the minimum bactericidal concentration (MBC). Gentamicin levels must be closely monitored because patients with endocarditis are particularly susceptible to the toxic effects as a result of renal impairment. Therapy is continued for 2–6 weeks depending on the MIC of the organism. Typical regimens include benzylpenicillin and gentamicin for viridans streptococci; flucloxacillin and gentamicin for staphylococci; vancomycin; gentamicin for penicillin-allergic patients. There are a number of national guidelines for management of endocarditis.

Surgical management may be required to deal with the haemodynamic consequences of endocarditis, especially in cases caused by *S. aureus* and other more virulent pathogens, or if infection is unresponsive to antimicrobial therapy.

Prevention

Endocarditis may be prevented by giving antibiotic prophylaxis to patients with damaged valves when they undergo procedures which give rise to significant bacteraemia, such as dental work or urogenital surgery. If the urine is infected, the antibiotic choice should reflect the sensitivity of the urinary organism cultured. For procedures requiring an anaesthetic, prophylaxis is given at induction followed by subsequent oral doses. There are alternative regimens for penicillin allergy and prosthetic valves laid down by national guidelines.

Myocarditis

Most myocarditis is caused by viral infection, of which enteroviruses are the most common cause. However, it may complicate systemic viral infections, follow bacteraemia or form part of brucellosis, rickettsial or chronic Chagas' infection.

Patients present with influenza-like symptoms associated with fatigue, exertional dyspnoea, palpitations and precordial pain. Tachycardia, dysrhythmia or cardiac failure may be present. The electrocardiogram (ECG) may show T-wave inversion, prolongation of the PR or QRS interval, extrasystoles or heart block. There may be an elevation in cardiac enzymes and cardiomegaly on chest X-ray.

The diagnosis is suggested by the relationship of viral symptoms to the development of cardiological abnormalities. Enteroviruses may be recovered from throat and faecal specimens, and respiratory viruses from nasopharyngeal or throat specimens (see Chapter 32) for culture or NAAT. Treatment is supportive.

Pericarditis

Pericarditis is most often secondary to a non-infectious condition, such as myocardial infarction. It may also arise as a complication of bacteraemia, following spread of pus from an empyema (*Streptococcus pneumoniae*), or from a liver abscess (enterococci, *Entamoeba histolytica*). Tuberculosis can cause subacute pericarditis.

Viral pericarditis is a self-limiting condition featuring fever, flu-like symptoms and sharp chest pain. Enteroviruses, especially coxsackie and influenza viruses, are most commonly implicated. The chest pain may vary with posture, swallowing or heartbeat. A pericardial rub may be heard. Cardiographic evidence of pericarditis may be demonstrated.

Patients with suppurative pericarditis present with fever, neutrophilia and signs of the underlying source of infection. Chest pain is severe and a fall in blood pressure may indicate developing tamponade. Electrocardiographic changes show upward-curved elevated ST segments. Echocardiography will show pericardial thickening or effusion. Infection can be complicated by fibrosis and constrictive pericarditis, leading to congestive cardiac failure. Treatment is directed against the likely causative organism.

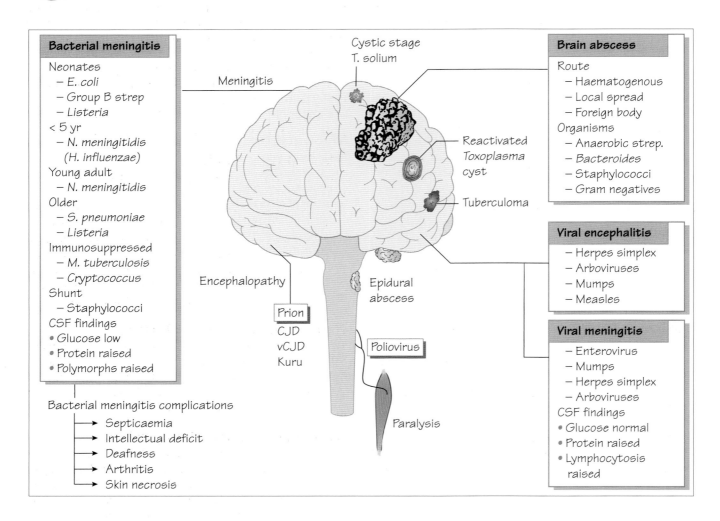

Bacterial meningitis
Aetiology
Many bacteria cause meningitis, with a different aetiological spectrum at different ages and in different patient groups (see figure).

Clinical features
Classically, meningitis presents with fever, headache, photophobia and neck stiffness. Occasionally vomiting and diarrhoea may predominate. The level of consciousness progressively falls. Not all these signs and symptoms may be present, especially in neonates and elderly people where presentation may be atypical. A bulging fontanelle in a neonate indicates raised intracranial pressure. Ear or sinus infections may suggest a pneumococcal cause.

Complications
Haemophilus influenzae meningitis can be complicated by recurrence of fever, hydrocephalus, convulsions and deafness. *Neisseria meningitidis* can be complicated by deafness, intellec-

tual deficit, skin necrosis and reactive arthritis. With appropriate treatment, mortality from *H. influenzae* and *N. meningitidis* should be less than 5%. However, *N. meningitidis* septicaemia can be rapidly fatal if the the diagnosis is delayed. Sequelae including deafness, cranial nerve palsies and hydrocephalus are most frequent following *Streptococcus pneumoniae* meningitis, which also has the highest mortality (> 20%).

Diagnosis
A sample of CSF should be obtained once raised intracranial pressure has been ruled out. The total and differential white cell count should be measured, and Gram stain, Ziehl-Neelsen, India ink, NAAT and antigen detection methods used. Bacterial meningitis causes a high white cell count, predominantly neutrophils, a low glucose and raised protein. In tuberculous meningitis the cells are mainly lymphocytes with high protein and low glucose. Blood should be taken for culture and rapid antigen detection, and glucose determination. Menincococcal disease can be diagnosed by NAAT on whole blood.

Management

Neonatal meningitis (likely organisms *Escherichia coli*, group B *Streptococcus* and *Listeria*) are treated with cefotaxime and an aminoglycoside, with ampicillin added if *Listeria* is suspected. *Haemophilus* meningitis requires cefotaxime. *Neisseria meningitidis* is invariably susceptible to penicillin. Penicillin-resistant *S. pneumoniae* is now being reported and cefotaxime can be substituted. If cephalosporin resistance is likely, vancomycin can be added. Cryptococcal meningitis is treated with amphotericin and 5-flucytosine. Tuberculous meningitis is treated with rifampicin, pyrazinamide, isonazid and ethambutol (see Chapter 17). Shunt-related meningitis should be treated according to the identity and susceptibilities of the organisms.

Prevention

Capsular polysaccharide vaccines are available for *N. meningitidis* serogroups A, C and W135 but not serogroup B which causes most cases in the UK. Conjugate vaccines against *H. influenzae*, *N. meningitidis* C are available and effective. Family (close) contacts of meningococcal and *Haemophilus* meningitis patients require antimicrobial prophylaxis (ciprofloxacin or rifampicin).

Brain abscess

Brain abscesses arise from parameningeal suppuration, foreign bodies or haematogenous spread from distant sepsis. Infection is polymicrobial with anaerobic cocci, *Prevotella* spp., staphylococci, streptococci (*S. anginosis-constellatis* group) and Enterobacteriaceae.

Clinical features

Patients present with headache, fever and reduced consciousness. Focal neurological signs depend on the location of the abscess. Signs of raised intracranial pressure may develop (rising blood pressure, falling pulse) followed by seizures.

Diagnosis and treatment

Lesions are localized by CT scanning. A lumbar puncture is contraindicated because of the risk of cerebellar herniation. Drainage should be performed if feasible and pus sent for culture and sensitivity testing.

In addition, a regimen of cefotaxime, metronidazole and penicillin or a combination of benzylpenicillin, chloramphenicol and metronidazole may be used.

Viral meningitis

Meningitis and encephalitis may arise from infection with enteroviruses, mumps, herpes simplex, arboviruses, influenza and, rarely, rubella or Epstein–Barr virus. Viral meningitis can be part of the natural history of polio infection (see Chapter 34). Patients present with headache, photophobia, fever and neck stiffness. The CSF shows an increase in lymphocytes; the protein is mildly raised with normal glucose levels. Throat swabs, CSF and stool specimens should be sent for viral culture and serological testing. Management is symptomatic as most patients recover without residual deficit within a few days.

Viral encephalitis

Viral encephalitis is caused by a variety of viruses including herpesvirus and arbovirus (see Chapters 29 and 36). Patients are febrile with headache, neck stiffness and impaired consciousness. Focal neurological signs may develop; convulsions are common. Virus may be cultured from CSF, stool and throat specimens, and detected by serological techniques. Aciclovir is used for treatment of herpetic encephalitis (which typically affects the temporal lobe) reducing both the mortality rate to less than 20%, and the number of patients with severe residual disability.

Postinfectious encephalitis

Some viruses are associated with encephalitis after a systemic infection has resolved (postinfectious encephalitis): measles, varicella zoster, rubella, Epstein–Barr virus, mumps and influenza. Clinically similar to viral meningitis, it is thought to be mediated by an autoimmune reaction.

Spongiform encephalopathies

The prion protein is a protease-resistant form of a protein that is a normal constituent of the brain. When ingested, the prion protein induces a conformational change in the host brain protein, leading to spongiform degeneration in the brain. There is an extended incubation period of more than 5 years.

Kuru was described in cannibals from Papua New Guinea who ate human tissue, including brain. Bovine spongiform encephalopathy (BSE) has arisen following the feeding of animal brain protein to cattle. Transmission to humans, following ingestion of contaminated beef products, is thought to be responsible for variant Creutzfeldt–Jakob disease. BSE in cattle has been eliminated in countries with a ban on feeding animal protein to cattle. The size and scale of any human epidemic is not yet known but appears to be small.

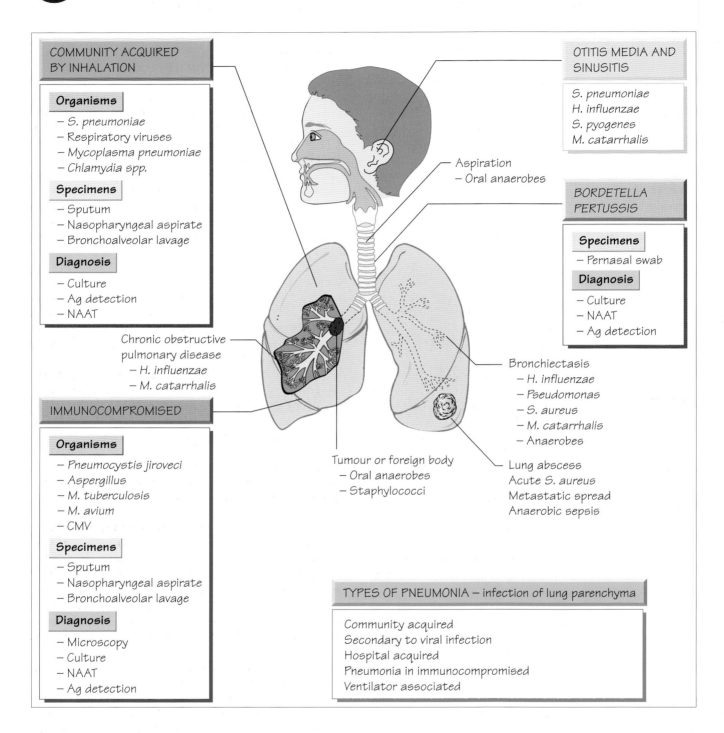

COMMUNITY ACQUIRED BY INHALATION

Organisms
- S. pneumoniae
- Respiratory viruses
- Mycoplasma pneumoniae
- Chlamydia spp.

Specimens
- Sputum
- Nasopharyngeal aspirate
- Bronchoalveolar lavage

Diagnosis
- Culture
- Ag detection
- NAAT

Chronic obstructive pulmonary disease
- H. influenzae
- M. catarrhalis

IMMUNOCOMPROMISED

Organisms
- Pneumocystis jiroveci
- Aspergillus
- M. tuberculosis
- M. avium
- CMV

Specimens
- Sputum
- Nasopharyngeal aspirate
- Bronchoalveolar lavage

Diagnosis
- Microscopy
- Culture
- NAAT
- Ag detection

OTITIS MEDIA AND SINUSITIS

S. pneumoniae
H. influenzae
S. pyogenes
M. catarrhalis

Aspiration
- Oral anaerobes

BORDETELLA PERTUSSIS

Specimens
- Pernasal swab

Diagnosis
- Culture
- NAAT
- Ag detection

Bronchiectasis
- H. influenzae
- Pseudomonas
- S. aureus
- M. catarrhalis
- Anaerobes

Tumour or foreign body
- Oral anaerobes
- Staphylococci

Lung abscess
Acute S. aureus
Metastatic spread
Anaerobic sepsis

TYPES OF PNEUMONIA – infection of lung parenchyma

Community acquired
Secondary to viral infection
Hospital acquired
Pneumonia in immunocompromised
Ventilator associated

Upper respiratory tract infections
Pharyngitis

This is a common condition in community practice, caused by viruses such as adenovirus and coxsackievirus, and bacteria. *Streptococcus pyogenes* is the most common bacterial cause, but *Neisseria gonorrhoeae* and *Candida* also cause pharyngitis.

Patients have fever and a painful infected throat that may have visible pus or exudate. Regional lymph nodes may be painful and enlarged. Streptococcal infection may be complicated by peritonsillar abscess (quinsy), bacteraemia, rheumatic fever or nephritis.

Corynebacterium diphtheriae infection should be considered if there is an appropriate travel history. If diphtheria is suspected, throat swabs should be taken and the laboratory alerted so that they will be inoculated onto appropriate media. Diphtheria can cause a green–black necrotic pharyngeal 'membrane' associated with neck oedema in severe cases.

The majority of infections in adults are viral so symptomatic treatment is adequate in many cases, but penicillin V or a macrolide may be given when bacterial infection is suspected or proved by near-patient testing. Ampicillin should be avoided as it provokes a rash with Epstein–Barr virus infection. Tonsillectomy and adenoidectomy may reduce the number of infective episodes of pharyngitis or otitis media in patients with quinsy or recurrent otitis media.

Otitis media and sinusitis

Infection occurs when sinuses or the middle ear are occluded by inflammation. Children under 7 years are especially prone to otitis media because the eustachian tube is short, narrow and nearly horizontal. The main infecting organisms are *Streptococcus pyogenes*, *Streptococcus pneumoniae*, *Haemophilus influenzae*, *Moraxella catarrhalis* and the more recently recognized *Aloiococcus otitidis*.

Patients present with fever and local intense pain. Small children may have difficulty in localizing the pain. In sinusitis, the pain is often worse with head movement and in the evening. Ear infection may be complicated by perforation, recurrent or chronic infection or the development of 'glue ear' (sterile mucus within the middle ear). Acute meningitis or mastoiditis complicates severe infection rarely.

Diagnosis is clinical; an auroscope reveals retrotympanic fluid levels, an inflamed tympanic membrane or a purulent discharge associated with perforation. Treatment depends on reducing mucosal swelling, promoting drainage of fluid and encouraging the recirculation of air. Appropriate antibiotic therapy has a role in this process.

Acute epiglottitis

This infection causes swelling of the epiglottis that may threaten the airway. *Haemophilus influenzae* type b was the most common cause until vaccination became available. Infection with *S. pyogenes* causes some cases, usually in adults. The presentation is with a sore throat and high fever, and often stridor and drooling. Examination of the throat should be avoided as it may precipitate acute respiratory obstruction. Treatment is with parenteral third-generation cephalosporins. Emergency tracheostomy may become necessary.

Lower respiratory tract infections

Lower respiratory tract infections (LRTIs) are an important cause of morbidity and mortality worldwide. They are the leading cause of death in children under the age of 5 years in developing countries.

Patients are predisposed to community-acquired pneumonia by factors including smoking, chronic obstructive pulmonary disease, diabetes mellitus, immunosuppressive therapy and HIV.

Many viruses cause primary viral pneumonia (e.g. influenza and SARS coronavirus). Others cause damage to the lower respiratory tract, permitting secondary bacterial pneumonia (see Chapter 33).

Clinical features

Patients have fever and a cough. Sputum may be purulent or blood-stained, although in some infections (e.g. *Mycoplasma*) productive cough is uncommon. Inflammation of the pleura causes sharp chest pain, worse on inspiration. Patients with LRTIs may also show signs of systemic infection, such as myalgia, malaise and weakness. In elderly people, mental confusion is common even when specific symptoms and signs are slight.

Pneumonia, especially with *S. pneumoniae*, can be complicated by local spread to the pleura and pericardium, and by septicaemia and meningitis. *Staphylococcus aureus* infection can be complicated by lung cavitation and bronchiectasis after recovery.

Diagnosis

Only sputum should be collected and physiotherapy may be helpful in obtaining a good-quality specimen. In patients too ill to produce a sputum specimen, bronchoalveolar lavage can be performed and is especially valuable for diagnosis of immuno-compromised patients.

Culture allows species identification and sensitivity testing. Antigen detection methods and NAATs are available for *Chlamydia*, *Mycoplasma*, *Legionella*, *Coxiella* and *S. pneumoniae* and respiratory viruses. NAAT can be performed quickly enough to inform treatment choice.

Management and prevention

Appropriate antibiotic therapy should be commenced as soon as possible. Severe community-acquired pneumonia requires hospitalization with intravenous antibiotics (e.g. a third-generation cephalosporin and macrolide). Milder infections can be treated orally, often with amoxicillin and/or a macrolide, although quinolones such as moxifloxacin are also used. As β-lactam resistance is common in *H. influenzae*, patients with chronic obstructive pulmonary disease should be treated with an appropriate agent (e.g. co-amoxiclav or trimethoprim). Treatment of hospital-acquired pneumonia may require agents active against Enterobacteriaceae and *Pseudomonas* (e.g. ciprofloxacin or ceftazidime).

Supportive therapy, including bed rest, oxygen, rehydration, physiotherapy and ventilation may be needed.

Infective exacerbations of cystic fibrosis are with *H. influenzae* initially, but infection with *Pseudomonas* and *Burkholderia cepacia* require specialist management with detailed culture and susceptibility testing that allows the optimization of antimicrobials. This should be coupled with intensive postural drainage and physiotherapy.

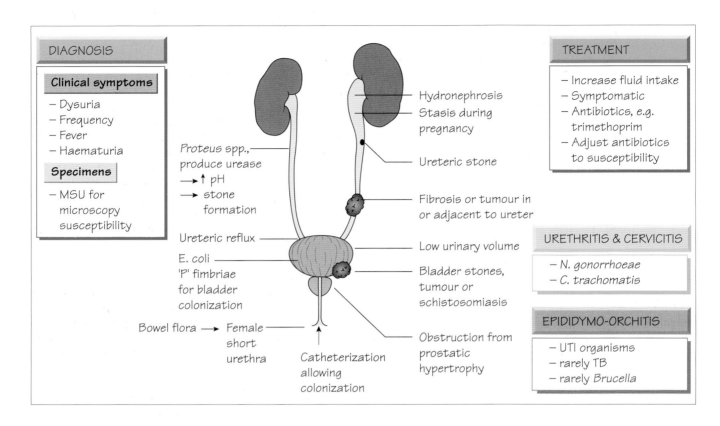

DIAGNOSIS

Clinical symptoms
- Dysuria
- Frequency
- Fever
- Haematuria

Specimens
- MSU for microscopy susceptibility

Proteus spp., produce urease
→↑ pH
→ stone formation

Ureteric reflux

E. coli 'P' fimbriae for bladder colonization

Bowel flora → Female short urethra

Catheterization allowing colonization

Hydronephrosis
Stasis during pregnancy

Ureteric stone

Fibrosis or tumour in or adjacent to ureter

Low urinary volume

Bladder stones, tumour or schistosomiasis

Obstruction from prostatic hypertrophy

TREATMENT
- Increase fluid intake
- Symptomatic
- Antibiotics, e.g. trimethoprim
- Adjust antibiotics to susceptibility

URETHRITIS & CERVICITIS
- N. gonorrhoeae
- C. trachomatis

EPIDIDYMO-ORCHITIS
- UTI organisms
- rarely TB
- rarely Brucella

Urinary tract infection
Anatomical considerations
Only the lower part of the urethra is usually colonized by bacteria; the flushing action of urinary flow protects against ascending infection. As the female urethra is short, urinary infection is more common in women.

Epidemiology and pathogenesis
Dehydration, obstruction, the disturbance of smooth urinary flow or the presence of a foreign body such as a stone or urinary catheter predisposes to urinary infection. Trauma during sexual intercourse may precipitate infection in women. Infection in children, especially in boys, is often associated with congenital abnormalities, such as ureteric reflux or urethral valves.

The most commonly isolated pathogens are *Escherichia coli*, *Klebsiella* spp. and *Enterococcus* spp. *Escherichia coli* uses fimbriae to adhere to the urinary epithelium, reducing the risk of being washed away. Infections caused by *Proteus* spp. are more likely in patients who have stones: *Proteus* spp. have urease activity that raises urinary pH, thus encouraging stone formation. *Staphylococcus saprophyticus* is a common isolate from sexually active females. Many different Gram-negative

organisms colonize urinary catheters, often producing invasive infections with bacteraemia.

Clinical features
Lower urinary tract infections are characterized initially by urinary frequency, dysuria and suprapubic discomfort; fever may be absent. In pyelonephritis, fever, loin pain, renal angle tenderness and signs of septicaemia may be present. In children, elderly people and antenatal patients, urinary infection may be clinically silent. Recurrent infections can result in scarring and renal failure.

Laboratory diagnosis
Urinary white blood cells and epithelial cells are used to assess specimen quality and significance of isolates. Urine can be contaminated by perineal organisms. This risk is minimized by taking a midstream urine (MSU) specimen and considering that $> 10^5$ c.f.u. per mL of a single organism indicate infection whereas $< 10^5$ organisms per mL or a mixed growth suggests contamination. This numerical approach is not always appropriate, however; chronically catheterized patients usually have 'significant' numbers of organisms and multiple pathogens. In

contrast, all isolates are potentially significant in a suprapubic aspirate from an infant with suspected infection. Susceptibility tests should be performed on all significant isolates.

Treatment and prevention

Antibiotic choice should be defined by susceptibility tests; empirical therapy should follow the known susceptibilities of urinary pathogens in that community. Most community-acquired infections respond to oral antibiotics, such as cefalexin, amoxicillin or trimethoprim. Should evidence of septicaemia be present, ciprofloxacin, cefotaxime or gentamicin may be required. Patients with recurrent urinary infection may require nocturnal prophylaxis (e.g. low-dose trimethoprim, nitrofurantoin or naladixic acid), together with advice on ensuring adequate urine flow. Children with recurrent infections should be investigated and may require surgical correction of anatomical abnormalities. Significant bacteriuria in pregnant women should be treated, even if asymptomatic. Anatomical obstructions to urine flow, such as stricture or stones, should be removed if possible. The risk of urinary tract infection is reduced by drinking enough fluids to ensure an adequate urinary flow.

Genital infection

Genital infection presents in many ways (Table 49.1). Other sites may be involved, for example the throat and rectum in gonococcal infection. It may be followed by pelvic inflammatory disease, infertility, prostatitis, arthritis or bacteraemia.

Diagnosis

Urethral and cervical swabs should be taken for both bacterial and viral diagnosis. *Neisseria gonorrhoeae*, *Chlamydia* and herpes simplex may be detected by culture methods but improved sensitivity is obtained by NAAT (see Chapter 26). Samples positive for *N. gonorrhoeae* can be cultured to permit susceptibility testing. Syphilis is diagnosed with EIA together with traditional treponemal tests (see Chapter 27). Direct microscopy may show evidence of *Candida* or *Trichomonas*.

Treatment

Chlamydia are responsive to fluoroquinolones, macrolides or tetracyclines. Penicillins are still the treatment of choice for gonorrhoea if sensitive, although cephalosporins, quinolones or spectinomycin are usually required. Syphilis is treated with penicillin (see Chapter 27). Patients must often be treated before a laboratory diagnosis is available and in developing countries treatment is guided by a 'syndromic approach' in which therapy is chosen to suit the susceptibilities of the likely pathogens. For example, patients with uncomplicated urethritis can be treated with a single dose of a suitable cephalosporin or fluoroquinolone followed by a 1-week course of either doxycycline or singledose azithromycin.

Prevention

Prevention requires risk avoidance (e.g. monogamous relationships) or risk reduction (e.g. barrier contraceptive methods). Sexual contacts of cases are traced to treat asymptomatic disease and reduce transmission. Antigen variability in *N. gonorrhoeae* means that there are no effective vaccines for gonorrhoea.

Trichomonas vaginalis

This protozoan causes an itchy vaginal infection, which presents as a discharge with an offensive smell. Treatment is with metronidazole. Treatment of sexual contacts may be necessary to prevent recurrent infection.

Non-specific vaginosis

This is caused by disruption to the normal vaginal flora. A mixture of organisms, including anaerobes, *Mobiluncus* spp. and *Gardnerella vaginalis*, may result in an offensive discharge with a characteristic fishy smell when alkalinized. The diagnosis is based on clinical findings and near-patient tests, for example the presence of epithelial cells heavily coated with bacteria in the discharge, and a positive amine test. Diagnosis is confirmed using defined syndrome scoring schemes. Non-specific vaginosis is treated with metronidazole.

Epididymo-orchitis

Infection of the epididymis may arise (a) from a urinary tract infection, (b) as part of a genital infection or (c) as a primary systemic infection, such as brucellosis or tuberculosis. Patients present with a painful, acutely inflamed epididymis and testis, which must be differentiated from testicular torsion. Diagnosis is made clinically and confirmed by the result of urinary or blood cultures and tests for sexually transmitted infections.

Table 49.1 Genitourinary infection syndromes and causative organisms.

Syndrome	Organisms
Genital ulcers	Herpes simplex *Chlamydia trachomatis* types L1–4 *Haemophilus ducreyi* (see Chapter 20) *Treponema pallidum* (see Chapter 27) *Calymmatobacterium donovani*
Urethral discharge	*Neisseria gonorrhoeae* *C. trachomatis*
Pelvic inflammatory disease	*N. gonorrhoeae* *C. trachomatis* Mixed anaerobic infection
Vaginal discharge	*Candida albicans* *Trichomonas vaginalis* *Mobiluncus* spp. and others in non-specific vaginitis

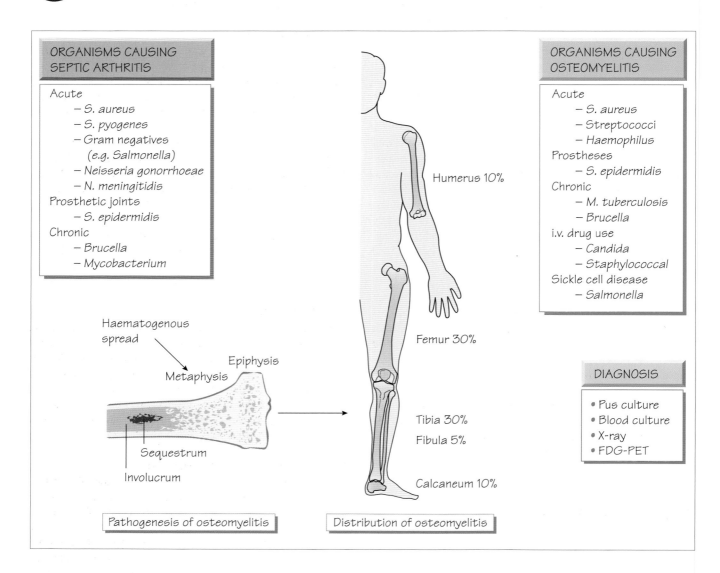

ORGANISMS CAUSING SEPTIC ARTHRITIS

Acute
- S. aureus
- S. pyogenes
- Gram negatives
  (e.g. Salmonella)
- Neisseria gonorrhoeae
- N. meningitidis
Prosthetic joints
- S. epidermidis
Chronic
- Brucella
- Mycobacterium

ORGANISMS CAUSING OSTEOMYELITIS

Acute
- S. aureus
- Streptococci
- Haemophilus
Prostheses
- S. epidermidis
Chronic
- M. tuberculosis
- Brucella
i.v. drug use
- Candida
- Staphylococcal
Sickle cell disease
- Salmonella

Humerus 10%

Femur 30%

Tibia 30%
Fibula 5%

Calcaneum 10%

Haematogenous spread

Epiphysis
Metaphysis

Sequestrum

Involucrum

Pathogenesis of osteomyelitis

Distribution of osteomyelitis

DIAGNOSIS
- Pus culture
- Blood culture
- X-ray
- FDG-PET

Osteomyelitis

Osteomyelitis (infection of bone) may arise from haematogenous spread, by direct extension from an infected joint, or following trauma, surgery or instrumentation. The formation of pus precipitates ischaemia and necrosis; the central area of dead bone is known as the sequestrum. New bone (the involucrum) may form around the infection site. In children, the metaphysis of the long bones (femur, tibia and humerus) are most often involved. In addition to these sites, infection of the spine is common in adults.

Staphylococcus aureus accounts for 90% of infections; rarer causes include *Streptococcus pyogenes* (4%), *Haemophilus influenzae* (4%), *Escherichia coli*, *Salmonella* spp., *Mycobacterium tuberculosis* and *Brucella*. Patients with sickle cell disease are especially prone to *Salmonella* infection.

Clinical features

Patients present with fever and pain. In some patients, especially the young, pain may be poorly localized. Young children may stop moving the affected limb (pseudoparalysis). As infection progresses, soft-tissue swelling may occur which may be followed by sinus formation. Pathological fractures may develop if diagnosis and treatment are delayed. Delayed treatment increases the risk of chronic osteomyelitis developing. Acute or chronic infection may develop around foreign bodies in the bone such as surgical nails or debris from trauma.

Diagnosis

Radiological changes do not develop until late in the course of infection when demineralization has occurred. Isotope scans may be helpful but do not distinguish infection from other inflammatory conditions. The most sensitive method of detecting osteomyelitis is fluorodeoxyglucose positron emission tomography (FDG-PET). It is essential that blood cultures are taken but these may be negative early in the course. Pus from bone via needle or open biopsy allows culture for pathogen identification and susceptibility testing.

Management

Drainage and excision of the sequestrum is an important part of the management. Empirical antibiotic therapy (e.g. flucloxacillin and fusidic acid) should be started at once, pending culture results. This choice is guided by the fact that staphylococci and streptococci are the commonest organisms found in community practice. Other agents such as ciprofloxacin may be required if, for example, *Salmonella* is isolated or suspected because the patient has sickle cell disease. Treatment lasts for 6 weeks or until there is evidence that inflammation has disappeared and the bone has healed.

Chronic osteomyelitis

Chronicity may arise from inadequately treated acute infection, or secondary to surgery or fracture. Infection of prosthetic materials with organisms with reduced virulence (coagulase-negative staphylococci) is increasingly common with the growth of prosthetic surgery in the ageing population. *Staphylococcus aureus* is implicated in 50% of cases; the remainder are associated with Gram-negative pathogens (*Pseudomonas*, *Proteus* and *E. coli*). Ongoing pain, swelling and deformity, with a chronically discharging sinus, are the main clinical features. A diagnosis by culture is essential; specimens should be taken under aseptic conditions. A prolonged course of appropriate antibiotics should accompany appropriate surgery. If there is an infected prosthetic device then this will usually need to be removed for treatment to be effective.

Suppurative arthritis

Suppurative arthritis usually arises from a bacteraemia; 95% of cases are caused by *S. aureus* and *S. pyogenes*. Other causes include Enterobacteriaceae, *Neisseria gonorrhoeae*, *H. influenzae*, *Salmonella* spp., *Brucella* spp., *Borrelia burgdorferi*, *Pasteurella* and *M. tuberculosis*. The large joints (e.g. the knee) are most commonly infected, but infection of the shoulder, hip, ankle, elbow and wrist joints may also occur. Prosthetic joints may become infected with skin contaminants (usually *S. aureus* or *Staphylococcus epidermidis*) at the time of operation, or from haematogenous spread. The original source will dictate the likely causative pathogen.

Clinical features

In children, the onset may be abrupt, with fever, pain and swelling of the joint associated with reduced movement. In adults, the onset may be insidious; a history of recent urinary infection or salmonellosis may be reported. Other associated signs include cellulitis or specific rashes, such as gonococcal skin rash.

Septic arthritis must be differentiated from acute rheumatoid arthritis, osteoarthritis, gout, pseudogout or reactive arthritis. A diagnostic tap will yield cloudy fluid, and Gram stain and white blood cell count may suggest infection that can be confirmed by culture within 48 h. Bone marrow for culture should be obtained when brucellosis is suspected.

Intravenous antibiotics, appropriate to the infecting organisms isolated or suspected, should be commenced, and oral therapy continued for up to 6 weeks. Aspiration and irrigation of the joint may be helpful in severe cases by reducing inflammatory damage.

Viral arthritis

Some viruses are associated with arthritis, for example rubella, mumps and hepatitis B. Rubella-related arthritis is more common in females and develops a few days after the rash. Several of the alphaviruses cause severe bone and joint symptoms. Arthritis caused by an immune response to the pathogen can follow recovery, for example after meningococcal disease, or *Shigella* or *Chlamydia* infection. The latter can be associated with uveitis and is known as Reiter's syndrome.

Prosthetic joint infections

Prosthetic joints may become infected at the time of operation or as a result of haematogenous spread. Organisms are often of low virulence, such as *S. epidermidis*. Infection with *S. aureus*, especially if methicillin resistant (MRSA), can have serious consequences. Treatment is with intravenous antibiotics, depending on the susceptibility of the infecting organisms. Infection usually results in loss of the prosthesis which must be removed at operation; it is important to prevent infection by effective control measures in the ward and theatre. Patients undergoing prosthetic joint surgery should receive antibiotic prophylaxis with an agent active against *S. aureus*.

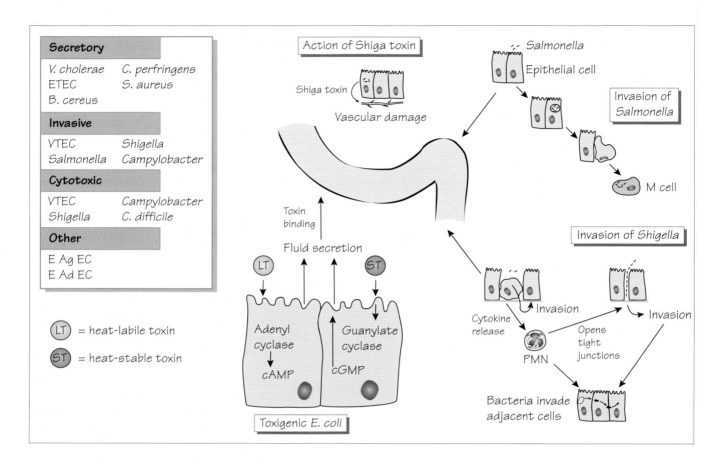

Infectious diarrhoea is a common condition, resulting in considerable economic loss from work absence. Worldwide it is one of the most important causes of death for children under 5. The gut is protected by gastric acid, bile salts, the mucosal immune system and inhibitory substances produced by the normal flora.

Organisms are transmitted by hands and fomites (faecal–oral route), by food or by water. The infective dose can be as few as 10 organisms (*Shigella*). Some foods (e.g. milk) or drugs (e.g. H_2 antagonists and proton pump inhibitors) may reduce the protective effects of gastric acid.

Bacteria can enter the food chain from infected animals, from poor hygiene during slaughter, and during butchering. Hens that are chronically colonized with *Salmonella* produce eggs that may be contaminated. Improper cooking and storage may allow the multiplication of bacteria (see below).

Transmission of diarrhoeal disease is also facilitated where there is poor sanitation, such as in general poverty, war or refugee crises. In these situations, infection spreads rapidly through the community, causing significant mortality. Cholera is capable of spreading worldwide (a pandemic).

Travellers' diarrhoea usually develops within 72 h of arrival in a new country; Latin America, Africa and Asia are the regions with the highest risk. Patients pass two to four watery bowel motions daily; blood and mucus are typically absent. The major organisms implicated are enterotoxigenic and enteroadherent *Escherichia coli*. Treatment is with fluid replacement and antibiotics, including co-trimoxazole or ciprofloxacin.

C. difficile-associated diarrhoea is a common hospital problem discussed in more detail in Chapter 18.

Pathogenesis

Infectious diarrhoea causes symptoms by a number of mechanisms, for example toxic deregulation of intestinal cells causing fluid secretion or invasion of the intestinal wall with destruction of the cells (see figure). Secretory diarrhoea produces infrequent large-volume stools as the absorptive capacity of the colon is overwhelmed. In dysenteric illness (*Shigella*), inflammation of the colon causes a loss in the capacitance, resulting in frequent stools that are often blood-stained. Enterohaemorrhagic *E. coli* (EHEC) produce Shiga toxin (Stx), causing bloody diarrhoea and the haemolytic–uraemic syndrome (HUS), a cause of renal failure in children. Serotype O157:H7 is the commonest. The microvascular endothelium is the toxin target.

Clinical features

Although diarrhoea may be defined as an increase in frequency of bowel action, it is a very subjective symptom. There may be many small stools (typical of large bowel infection), or infrequent large stools (small intestine infection). Stools may be blood-stained when there is destruction of the intestinal mucosa, or have a fatty consistency and offensive smell if malabsorption is present.

Dehydration and electrolyte imbalance may develop rapidly with potentially fatal results, as in cholera. Crampy abdominal pain may accompany diarrhoea (e.g. *Campylobacter* and *Shigella* infections); this may mimic acute abdominal conditions, such as appendicitis. Fever is not always present in diarrhoeal disease.

Septicaemia may develop in some cases of salmonellosis but is rare in other diarrhoeal diseases. Self-limiting bacteraemia is common in *Campylobacter* infection. Enterotoxigenic *Escherichia coli* O157 infection can produce a haemorrhagic colitis that is later complicated by renal failure and the haemolytic–uraemic syndrome. Secondary lactose intolerance resulting in continuing diarrhoea is caused by loss of intestinal lactase. It usually lasts a few weeks before resolving spontaneously. Patients with immunodeficiency may have difficulty eradicating intestinal infections: IgA deficiency, *Giardia*; T-cell deficiency, *Salmonella* and *Cryptosporidium* (see Chapter 53). Diarrhoea from viruses or protozoa is discussed in more detail in Chapters 34 and 39, respectively.

Diagnosis

Stool should be routinely examined microscopically for intestinal protozoa (e.g. *Giardia lamblia*). Ziehl–Neelsen's stain can be used to detect microsporidia and *Cryptosporidium parvum* (see Chapter 39).

Selective media must be used to culture bacterial pathogens so that the growth of non-pathogenic comensals is suppressed, for example sorbitol MacConkey for verotoxic *E. coli* (O157). Media can be made selective for *Campylobacter* by incorporating antibiotics and/or by incubating the plates at 43 °C. If cholera is suspected, stools are inoculated into alkaline peptone water (high pH allows *Vibrio cholerae* to grow preferentially); it can then be subcultured onto special selective medium containing bile salts and a high pH.

Multiplex systems of NAATs for bacterial diagnosis have been developed but have not yet entered routine practice. Organisms may be serotyped for epidemiological purposes. When organisms have only one serotype (e.g. *Shigella sonnei*), further typing (molecular typing) is required to confirm an outbreak.

The presence of viruses in stool can be demonstrated directly by electron microscopy, culture, EIA or NAATs (see Chapter 34). Toxin may be detected in stool samples, e.g. *Clostridium difficile* toxin.

Management

The management of diarrhoeal disease is based on adequate fluid replacement and restoration of electrolyte imbalances. Despite the outflow found in secretory diarrhoea, fluid absorption still occurs. Oral rehydration solutions consist of 150–155 mmol/L sodium and 200–220 mmol/L glucose, and can be life-saving. Intravenous fluid replacement is rarely necessary. Antimotility drugs are of no benefit and may be dangerous, especially in small children. Cholera and severe fluid diarrhoea may benefit from oral antibiotics, such as tetracycline or ciprofloxacin, which may shorten the duration of symptoms. Patients with severe dysentery and salmonellosis should be treated with ciprofloxacin or co-trimoxazole. Renal failure associated with haemolytic–uraemic syndrome following *E. coli* O157 requires specialist management.

Prevention

Water supplies, uncontaminated by human or animal faeces, are essential in preventing diarrhoeal disease. Animal husbandry and slaughter methods should be designed to prevent the introduction of animal intestinal pathogens into the human food chain. Food must be cooked to a sufficiently high temperature to kill pathogens and, if not eaten immediately, refrigerated at a low enough temperature to prevent bacterial multiplication.

Cooked food should be physically separated from uncooked to prevent cross-contamination. This is especially true in institutional cooking (e.g. hospitals and restaurants), where many may become infected following a single failure of hygiene.

Travellers' diarrhoea can be prevented by careful choice of food while travelling.

There are oral heat-killed and live attenuated cholera vaccines licensed for use but protection is short-lived. There are three parenteral inactivated whole-cell vaccines against typhoid: heat-inactivated phenol-preserved, acetone-inactivated dried, and purified Vi polysaccharide, as well as the oral Ty21a vaccine. New genetically engineered vaccines are being developed: a parenteral Vi antigen vaccine shows good immunogenicity in field studies and engineered live attenuated vaccines are in early trial.

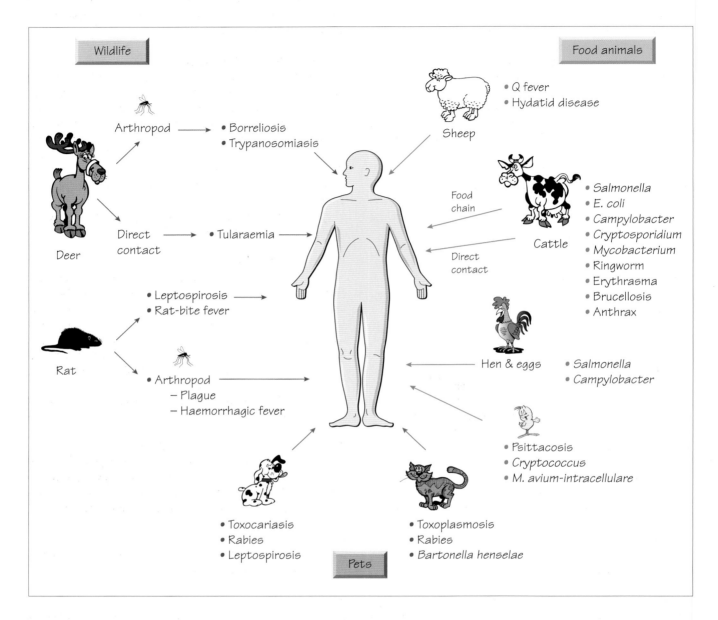

A zoonosis is an infection acquired from an animal source. Infections are acquired when humans enter the environment where the natural life-cycle occurs, for example when camping. Transmission can occur via vectors such as mosquitoes (e.g. Japanese B encephalitis). Alternatively, farming may expose the workers to infections from livestock. Pets are an important source of infection.

Viral zoonoses

More than 100 animal viruses can cause human disease; for example, herpes simiae, a monkey pathogen, causing severe encephalitis, or avian influenza, both causing a high mortality. Other viral zoonoses are discussed in Chapter 36.

Rat-bite fever

Rat-bite fever is caused by either *Streptobacillus moniliformis* or *Spirillum minus* inoculated by the bite of a rat. Following a 2-week incubation period, an inflammatory reaction is found at the site of the bite with lymphangitis and regional lymphadenopathy. There is a generalized maculopapular rash, together with fever, headache and malaise. Endocarditis is the most serious complication. Spontaneous recovery may occur within 2 months, but in untreated patients mortality is around 10%.

Diagnosis relies on visualizing the organism in tissue, bacterial isolation or NAAT. Treatment is usually with penicillin.

Anthrax

See Chapter 16.

Plague

Caused by *Yersinia pestis*, the infection is endemic in rodents in remote rural areas. Rarely, epidemics may develop which may spread worldwide (e.g. the Black Death). The organism is transmitted between rats, and to humans, by the rat flea, *Xenopsylla cheopis*.

The incubation period is short: the disease has an abrupt onset characterized by fevers and toxaemia. The regional lymph glands draining the site of the bite become greatly enlarged (buboes) and septicaemia is accompanied by generalized haemorrhage. Pneumonic plague is a rapidly fatal pneumonitis that can be transmitted by the respiratory route.

Plague is diagnosed clinically in areas where it is endemic. Direct smear of lymph gland aspirate or blood culture and NAAT is used for diagnosis. Treatment is with tetracycline, chloramphenicol, aminoglycosides or ciprofloxacin. The mortality rate of pneumonic plague is high. There are concerns that it may be used as a bioterrorism weapon.

Borreliosis

Borreliosis is transmitted from rodents or deer by ticks (open forest habitat, e.g. the New Forest) or by lice (see Chapter 27).

Toxoplasmosis

The cat is the definitive host of *Toxoplasma gondii* but the organism infects a wide range of animals, including sheep, cattle and humans. Infection is acquired by ingestion of oocysts from infected cat faeces or from tissue cysts in infected meat (e.g. undercooked beef).

Dermatophytes

Dermatophytes that are natural pathogens of animals can spread to the human population by direct contact (see Chapter 38).

Toxocariasis

Toxocara canis is an ascarid parasite of dogs. The parasite eggs are excreted in the faeces of infected dogs and mature in the soil. Human ingestion occurs when food is contaminated by the soil, or when personal hygiene is poor (e.g. hand-washing). The larval stages hatch in the intestine, invade the host and migrate to the liver and lungs. They are unable to develop into adults but migrate throughout the body causing fever, hepatosplenomegaly, lymphadenopathy and wheeze. If the larva migrates into the eye, sight may be permanently damaged by local inflammatory response of the retina. The diagnosis is made serologically using a specific EIA. The disease is usually self-limiting but, if symptoms are severe, treatment with albendazole may be beneficial. Ocular lesions should be treated first with steroids to diminish the inflammatory response; the role of anti-helminthic treatment is less certain.

Cat-scratch disease

Ten days following a cat scratch or bite, a papular lesion caused by *Bartonella henselae* may develop at the site. It is associated with regional lymphadenopathy. The symptoms resolve slowly over a period of 2 months, but a more chronic course may ensue. Cat-scratch disease can be complicated by disseminated infection; this is more common in immunocompromised individuals. Diagnosis is usually made clinically but can be confirmed serologically by immunofluorescence or EIA. Culture requires a prolonged incubation period and NAAT may also be used for diagnosis. Treatment with azithromycin, tetracyclines or rifampicin may be beneficial.

Hydatid disease

Two species of parasite are responsible for human hydatid disease: *Echinococcus granulosus* and *E. multilocularis*. Dogs are the definitive host of *E. granulosus*, harbouring the tapeworm stage; the eggs are passed in the faeces. The eggs are ingested by the intermediate hosts, e.g. sheep or rodents, and multiple cysts develop in the liver and lungs. The cycle is complete when dogs eat infected tissues. Humans are accidental hosts. The disease is common in sheep-farming areas. *Echinococcus multilocularis* is found in foxes, wolves and dogs; rodents act as the intermediate hosts.

Pathogenesis and clinical features

Cysts act as space-occupying lesions in the liver, lungs, abdominal cavity or central nervous system, and are responsible for the symptoms and signs of disease. The cysts of *E. multilocularis* lack a definite cyst wall and may ramify widely in the tissue .

Diagnosis

Cysts may be demonstrated by ultrasound or CT. EIA for both antibody and antigen is available.

Treatment

If possible, hydatid cysts should be surgically removed. Albendazole is given to kill the germinal layer of the cyst, and praziquantel to reduce the viability of protoscolices. Puncture, aspiration, injection of chemicals and re-aspiration (PAIR) is seen as an alternative to surgical excision. Cyst rupture can lead to multiple cysts in the abdomen, or anaphylaxis.

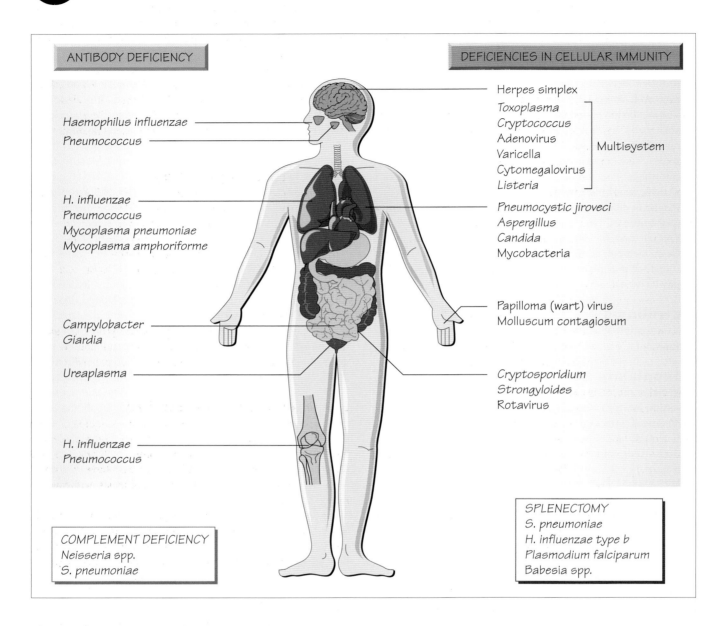

ANTIBODY DEFICIENCY

Haemophilus influenzae
Pneumococcus

H. influenzae
Pneumococcus
Mycoplasma pneumoniae
Mycoplasma amphoriforme

Campylobacter
Giardia

Ureaplasma

H. influenzae
Pneumococcus

COMPLEMENT DEFICIENCY
Neisseria spp.
S. pneumoniae

DEFICIENCIES IN CELLULAR IMMUNITY

Herpes simplex
Toxoplasma
Cryptococcus
Adenovirus   Multisystem
Varicella
Cytomegalovirus
Listeria

Pneumocystic jiroveci
Aspergillus
Candida
Mycobacteria

Papilloma (wart) virus
Molluscum contagiosum

Cryptosporidium
Strongyloides
Rotavirus

SPLENECTOMY
S. pneumoniae
H. influenzae type b
Plasmodium falciparum
Babesia spp.

Medical treatment or hereditary deficiency of components of the immune system may allow organisms with reduced virulence to cause infection and normal pathogens to cause severe infection. The origin of deficiency is often multifactorial; for example patients undergoing bone-marrow transplantation are neutropenic reducing resistance to bacterial infection, whilst intravenous cannulation provides a route for *Staphylococcus epidermidis* infection.

Medical treatment often breaches the physical barriers to infection (e.g. cannulation; see Chapter 9). Infections specifically associated with AIDS are discussed in Chapter 44.

Neutropenia
Granulocytopenia most often arises as a result of acute leukaemia or its treatment. The risk of infection depends on both the duration and severity of the neutropenia. Bacteraemia occurs in between 40 and 70% of neutropenic patients. The Enterobacteriaceae and *Pseudomonas* spp. are the most common Gram-negative bacilli isolated. These bacteria invade following gut damage by antineoplastic agents or irradiation. Gram-positive organisms (*S. epidermidis*, *S. mitis* and *S. oralis*, *Enterococcus* spp., *S. aureus* and *Corynebacterium jeikeium*) are increasingly important causes of sepsis.

Although antibiotic therapy may predispose to colonization by *Candida albicans*, fungal infection may occur *de novo* in the neutropenic patient. Increasingly, infections with yeasts such as *Candida krusei* (naturally resistant to antifungal therapy), *Aspergillus* spp. (causing invasive disease), and *Fusarium* spp., *Pseudallescheria boydii* and *Trichosporon beigelii* are being reported.

Treatment of fever in neutropenic patients

Empirical therapy includes carbapenem or ceftazidime and amikacin. If fever is unresolved, a glycopeptide can be added. Later, if fever still persists, fungal pathogens are more likely and amphotericin or itraconazole may be added.

Prevention of infection

The risk of infection in neutropenic patients is reduced when the patient is nursed in a side-room and supplied with sterilized water and food. Sterile procedures such as thorough hand-washing and latex gloves, should be employed; attendants should also wear gowns and masks. Room air is filtered to remove fungal spores.

Oral nystatin, alone or in combination with oral amphotericin, reduces the incidence of fungal infection. Fluconazole or itraconazole may also be useful. Antibiotic prophylaxis using 4-fluoroquinolones which targets the facultative anaerobes of the gut, preserving the anaerobic flora, is used in some centres.

T-cell deficiency

T-cell deficiency is an increasingly common problem following HIV infection, cancer chemotherapy, corticosteroid therapy or organ transplantation. Congenital T-cell deficiencies are rare but may be purely linked to T-cell function or combined with a hypogammaglobulinaemia.

Pathogens

These are mainly those microorganisms that have an intracellular location in the human host, such as:
- *Toxoplasma gondii, Strongyloides stercoralis*
- *Mycobacterium tuberculosis, M. avium-intracellulare*
- *Listeria monocytogenes, Cryptococcus neoformans, Pneumocystis jiroveci*
- herpes simplex, cytomegalovirus, varicella zoster virus and measles.

Measles infection, complicated by giant cell pneumonia and encephalitis, can be life-threatening.

Diagnosis

Specific infections should be investigated appropriately (see relevant chapters). All patients should have at least two blood cultures taken from different sites.

Hypogammaglobulinaemia

X-linked agammaglobulinaemia patients are at increased risk of infection for the first 6 months of life; the common variable immunodeficiency patients are at increased risk throughout life. Functional hypogammaglobulinaemia develops in patients with multiple myeloma.

Patients suffer recurrent respiratory tract infections with *Streptococcus pneumoniae*, the novel *Mycoplasma amphoriforme* and non-capsulate *Haemophilus influenzae,* leading to bronchiectasis. *Giardia, Cryptosporidium* and *Campylobacter* infections may be more persistent. Intravenous immunoglobulin reduces recurrent infection.

Complement deficiency

Hereditary complement deficiencies are rare. Deficiency in the later components of the complement cascade (C7–9) results in an inability to lyse Gram-negative bacteria, and patients are susceptible to recurrent *Neisseria* infection. Deficiency of the alternative complement pathway leads to serious *S. pneumoniae* infections, including meningitis. Acquired complement deficiency occurs in systemic lupus erythematosus.

Mannose-binding lectin

A wide range of bacteria fungi, viruses and protozoa bind to mannose-binding lectin and there are reports that infection with these organisms are more common or severe with some deficiency genotypes.

Postsplenectomy infection

The incidence of serious sepsis following splenectomy is about 1% per year; the rate is higher in infants and children. Highest mortality is associated with splenectomy for lymphoma and thalassaemia. Patients with sickle cell disease have functional asplenia. Although the risk of sepsis diminishes with time, it never disappears.

Streptococcus pneumoniae is responsible for approximately two-thirds of infections in most series; other important bacteria are *H. influenzae* and *E. coli*. Malaria may run a fulminant course. Splenectomy predisposes to *Capnocytophaga canimorsis* infection, usually arising after a dog bite.

Prevention

Vaccination against *S. pneumoniae* meningococcus and *H. influenzae* type b should be offered to all splenectomized patients. Improved responses may now be obtained with protein conjugate vaccines. Low-dose oral antibiotic prophylaxis with penicillin V should also be offered. Patients must be aware of the need to consult their doctor at the onset of any fever, or be instructed in the use of antibiotics, prescribed in advance to avoid delay in initiation of treatment.

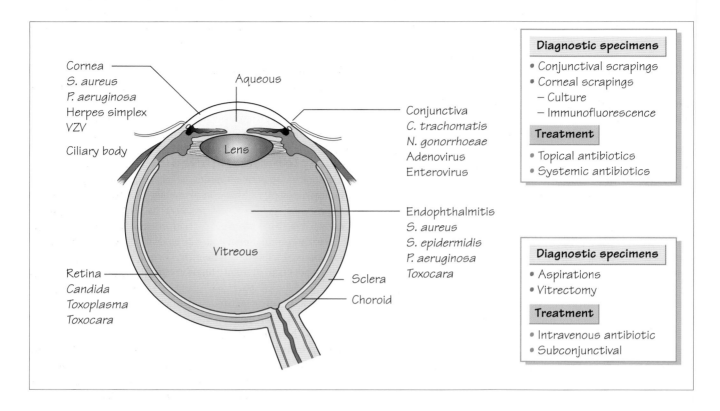

Cornea
S. aureus
P. aeruginosa
Herpes simplex
VZV

Ciliary body

Aqueous

Lens

Vitreous

Retina
Candida
Toxoplasma
Toxocara

Conjunctiva
C. trachomatis
N. gonorrhoeae
Adenovirus
Enterovirus

Endophthalmitis
S. aureus
S. epidermidis
P. aeruginosa
Toxocara

Sclera

Choroid

Diagnostic specimens
- Conjunctival scrapings
- Corneal scrapings
  - Culture
  - Immunofluorescence

Treatment
- Topical antibiotics
- Systemic antibiotics

Diagnostic specimens
- Aspirations
- Vitrectomy

Treatment
- Intravenous antibiotic
- Subconjunctival

Bacterial conjunctivitis

Bacterial conjunctivitis is a common condition caused by *Staphylococcus aureus*, *Haemophilus influenzae*, *Streptococcus pneumoniae* or *Moraxella* spp. Neonatal conjunctivitis may be caused by *Neisseria gonorrhoeae*, *Chlamydia trachomatis*, *Escherichia coli*, *S. aureus* and *H. influenzae* and is acquired from infection in the mother's genital tract. Infection with *Pseudomonas aeruginosa* can be acquired in hospital if ocular equipment or drops are not adequately sterilized or restricted to single use. Infection is also associated with contaminated personal contact-lens cleaning equipment. It produces a rapidly progressive infection that can result in ocular perforation and loss of vision. Irrespective of its aetiology, bacterial conjunctivitis presents with hyperaemic red conjunctivae and a profuse mucopurulent discharge. Conjunctival swabs and corneal scrapings are submitted for laboratory examination. The diagnosis is confirmed by bacterial culture or NAAT; *Chlamydia trachomatis* by NAAT. Treatment is by local antibiotics, including fusidic acid, tetracycline or chloramphenicol.

Adenovirus infection

Serotypes 7, 3, 10, 4 and 8 are the most common serotypes associated with ocular infection. Infection causes a purulent conjunctivitis, with enlargement of the ipsilateral periauricular lymph node. Half of patients with corneal involvement develop

punctate keratitis followed by subepithelial inflammatory infiltration. Anterior uveitis and conjunctival haemorrhages may develop. Treatment is symptomatic, with antibacterial agents being used if there is evidence of secondary bacterial infection. Topical steroids should be avoided.

Varicella zoster virus

The ophthalmic dermatome of the fifth cranial nerve is involved in approximately 10% of recurrent varicella zoster virus (VZV) infections (shingles). Ocular involvement, associated with lesions present on the skin of the tip of the nose, includes anterior uveitis, keratitis, ocular perforation or retinal involvement. Chronic disease occurs in about one-quarter of patients. The condition is very painful and may continue after healing of the rash (postherpetic neuralgia). Antiviral agents (e.g. aciclovir) should be used early in the infection and may prevent complications. Severe inflammation may benefit from topical steroids. A live attenuated vaccine is available to prevent primary infection.

Herpes simplex

Ocular infection with Herpes simplex is the most common infectious cause of blindness in developed countries. Typically, it presents with ulcerative blepharitis, follicular conjunctivitis and regional lymphadenopathy. Most patients have corneal involvement. Relapses occur approximately every 4 years. Initially, the dendritic ulcer is the marker of infection, but the later clinical picture is dominated by inflammation in deeper tissues, keratitis, corneal oedema and opacity. Primary infection and early relapses are treated with topical aciclovir. Inappropriate use of steroids worsens the keratitis. Progressive scarring, following repeated attacks, leads to corneal opacity and is one of the most common indications for corneal grafting.

Ocular manifestations of AIDS

'Cotton wool spots' are a common retinal manifestation of HIV infection. They follow infarction of the retinal nerve fibre layer, and may result in poor colour sensitivity and perception. Late in the course of HIV disease, especially before the introduction of HAART, when the CD4 count has fallen to below 0.05×10^9/L, ocular infection with cytomegalovirus may develop in up to one-third of patients. This causes a slowly progressive retinitis characterized by necrosis, and is an important cause of blindness in this patient group. The syndrome is difficult to differentiate from ocular toxoplasmosis or syphilitic retinitis. Initially, treatment with antiviral agents (e.g. ganciclovir) is given intravenously to control the disease; weekly maintenance therapy is required to prevent relapse once control is established.

Trachoma

Trachoma is a chronic keratoconjunctivitis caused by infection with *Chlamydia trachomatis*. It was once endemic throughout the world but is now largely confined to the tropics, where poor social conditions make transmission easier and poverty precludes adequate medical treatment. Symptoms develop 3–10 days after infection, with lacrimation, mucopurulent discharge, conjunctival infection and follicular hypertrophy. Treatment is with oral macrolides, such as azithromycin. An international campaign plan to eradicate trachoma by 2020 is under way using the SAFE strategy (surgery for inturned lids, antibiotics, face washing and environmental improvement).

Endophthalmitis

Endophthalmitis develops after ocular operation, following trauma and direct inoculation of a foreign body, and as a complication of systemic infection. Early postoperative infections are usually with *S. aureus*, *Staphylococcus epidermidis*, streptococci and Gram-negative bacilli. Late postoperative infections are with more indolent bacteria derived from the skin or acute infections caused by streptococci or *H. influenzae*. Post-traumatic infections are with *S. epidermidis*, *Bacillus* and streptococci. Endogenous infections secondary to bacteraemia or fungaemia are most often with *Candida*, streptococci and enteric Gram-negative bacilli. Rarely, endophthalmitis is caused by the nematode *Toxocara canis* (see Chapter 52).

Diagnosis is achieved by taking vitreous aspiration or vitrectomy specimens. Bacterial endophthalmitis is best managed using early vitrectomy and systemic antibiotics.

Onchocerciasis

Onchocerciasis is one of the most important causes of blindness in the world. It is caused by the filarial parasite *Onchocerca volvulus*. Heavy infection causes inflammatory lesions in the eye, which result in blindness. This parasite is discussed in more detail in Chapter 42.

55 Infections of the skin and soft tissue

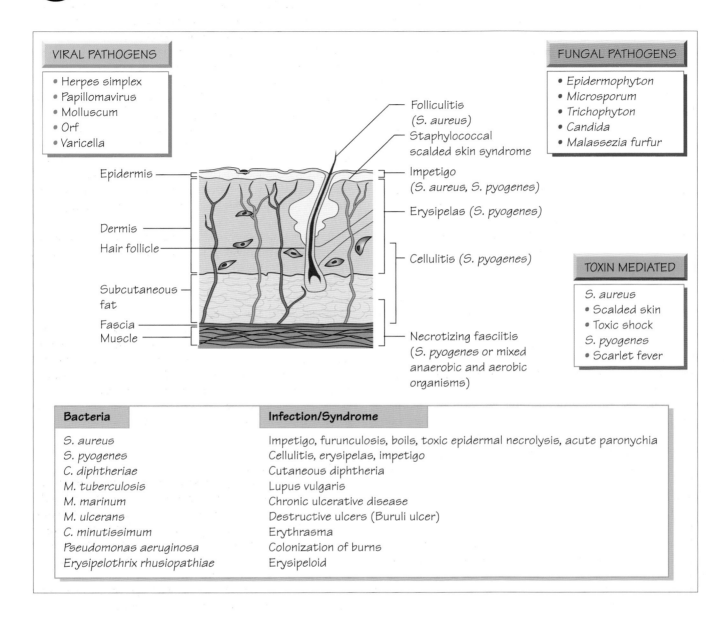

VIRAL PATHOGENS

- Herpes simplex
- Papillomavirus
- Molluscum
- Orf
- Varicella

FUNGAL PATHOGENS

- Epidermophyton
- Microsporum
- Trichophyton
- Candida
- Malassezia furfur

Epidermis

Dermis

Hair follicle

Subcutaneous fat

Fascia

Muscle

Folliculitis (S. aureus)

Staphylococcal scalded skin syndrome

Impetigo (S. aureus, S. pyogenes)

Erysipelas (S. pyogenes)

Cellulitis (S. pyogenes)

Necrotizing fasciitis (S. pyogenes or mixed anaerobic and aerobic organisms)

TOXIN MEDIATED

S. aureus
- Scalded skin
- Toxic shock

S. pyogenes
- Scarlet fever

Bacteria	Infection/Syndrome
S. aureus	Impetigo, furunculosis, boils, toxic epidermal necrolysis, acute paronychia
S. pyogenes	Cellulitis, erysipelas, impetigo
C. diphtheriae	Cutaneous diphtheria
M. tuberculosis	Lupus vulgaris
M. marinum	Chronic ulcerative disease
M. ulcerans	Destructive ulcers (Buruli ulcer)
C. minutissimum	Erythrasma
Pseudomonas aeruginosa	Colonization of burns
Erysipelothrix rhusiopathiae	Erysipeloid

Bacterial

Skin infections spread rapidly by contact, especially in enclosed populations or where sanitation is poor. A wide range of organisms infects the skin (see figure): *Staphylococcus aureus* and *Streptococcus pyogenes* are most commonly implicated.

Cellulitis affects all layers of the skin and can be caused by *S. pyogenes*, *S. aureus*, *Pasteurella multocida*, or rarely, marine vibrios or Gram-negative bacilli. Organisms invade via skin abrasions, insect bites or wounds. Empirical flucloxacillin should be given until culture results are available. Severe disease should be treated with intravenous antibiotics, including benzylpenicillin and flucloxacillin.

Necrotizing fasciitis is a rapidly progressive infection that spreads to involve skin and subcutaneous layers. Mixed aerobic and anaerobic infection or pure *S. pyogenes* infection may be responsible. It progresses rapidly and leads to death in a very short time. Effective treatment depends on adequate surgical resection of infected tissue, supplemented with benzylpenicillin, a third-generation cephalosporin and metronidazole.

Erythrasma is a superficial infection of the flexures caused by *Corynebacterium minutissimum*. Its lesions fluoresce under Wood's light. The organism may be cultured, and treatment is with erythromycin or tetracycline.

Erysipelas is a demarcated streptococcal infection confined to the dermis, usually found on the face or shins, which feels hot and looks red on examination. There is a modest increase in the peripheral white blood cells and fever may be present. Treatment with oral amoxicillin or flucloxacillin is usually effective but intravenous therapy may be required for severe cases.

Erysipeloid, a dull-red lesion, is a zoonosis caused by *Erysipelothrix rhusiopathiae*. Infection is usually acquired from inoculation injuries, usually in pig-handlers, butchers and fishermen. Disease may be self-limiting but treatment with oral penicillin or tetracycline speeds the response and is needed in rare cases of septicaemia.

Burns are very susceptible to colonization by bacteria; *Pseudomonas aeruginosa*, *S. aureus* and *S. pyogenes* and occasionally coliforms are implicated. Colonization with resistant organisms is an increasing problem. Bacterial colonization can lead to the loss of skin grafts and to secondary bacteraemia.

Paronychia

This is a common infection in community practice. The cuticle is damaged, allowing invasion with organisms such as *S. aureus*. There is pain and swelling, followed by a small abscess. The abscess may be drained and antibiotics given (e.g. flucloxacillin).

Manifestations of systemic infections

The skin is a large organ that can act as a window onto systemic infection. Examples include the petechial rash of meningococcal septicaemia heralding overwhelming sepsis. Patients with *Pseudomonas* septicaemia may have a gangrenous skin lesion, ecthyma gangrenosum. More subtle are the changes associated with endocarditis, such as splinter haemorrhages. Staphylococcal septicaemia may have evidence of skin infarctions. Many viruses cause lesions in the skin as part of a systemic infection (chickenpox and measles). In Herpes simplex infection, the skin is the primary site of infection (see Chapter 30). *Staphylococcus aureus* and β-haemolytic streptococci can both cause toxin-mediated systemic disease associated with different skin manifestations: toxic shock syndrome—generalized and palmar rash; scarlet fever—rash with circumoral pallor; scalded skin—desquamation in neonates.

Warts

Human papillomaviruses infect skin cells causing increased skin replication and giving rise to a wart. Papular, macular or mosaic variations can occur; verrucae (plantar warts) are found on the soles of the feet. The virus is transmitted by direct contact, par-ticularly under wet conditions, such as around swimming pools. Genital warts (condylomata acuminata) may be transmitted sexually. The diagnosis of warts is usually made clinically; virus in condylomata acuminata can be detected by immunofluorescence and PCR-based amplification techniques.

Papillomaviruses are associated with malignancy: cervical (HPV-16 and HPV-18); and laryngeal (HPV-6 and HPV-11). If warts are noted on routine cervical cytology, this is an indication for more frequent review. A vaccine has been developed and provides sustained immunity against HPV-16 and HPV-18.

Except in the immunocompromised, warts are self-limiting and spontaneously resolve without scarring. Topical keratolytic agents, e.g. salicylic acid, are widely available over the counter for self-application. Genital warts may respond to the application of podophyllum by trained staff. Cryotherapy is the second-line therapy which may hasten their departure. Cautery is no longer recommended.

Several poxviruses infect the skin and cause characteristic lesions, for example molluscum contagiosum and orf. These are discussed in Chapter 31.

Dermatophytes
Clinical features

Dermatophyte infection (ringworm) may present as itchy, red, scaly, patch-like lesions which spread outwards leaving a pale, healed centre. Chronic nail infection produces discoloration and thickening, whereas scalp infection is often associated with hair loss and scarring. Clinical diagnostic labels are based on the site of infection, e.g. tinea capitis (head and scalp), tinea corporis (trunk lesion). The causative organism may vary (see Chapter 38).

Laboratory diagnosis

Infection of skin and hair by some species may demonstrate a characteristic fluorescence when examined under Wood's light.

Skin scrapings, nail clippings and hair samples should be sent dry to the laboratory. Typical branching hyphal elements may be demonstrated under the microscopy in a potassium hydroxide preparation. Dermatophytes take up to 4 weeks at 30°C to grow on Sabouraud's dextrose agar.

Identification is based on colonial morphology, microscopic appearance (lactophenol blue mount), biochemical tests and sequencing of the 18S rRNA gene.

Treatment

Dermatophyte infections may be treated topically with imidazoles, e.g. miconazole, clotrimazole, tioconazole or amorolfine. Some infections require oral terbinafine for several weeks.

Self-assessment case studies: questions

Case 1

A 26-year-old male presented to the accident and emergency department of a large city hospital. He complained that he felt weak and unwell, 'as if I have flu', and that this had been getting worse over the last few days. The patient admitted to taking illicit drugs intravenously, and this was confirmed on examination with evidence of multiple puncture marks. He was hot to the touch and his skin appeared clammy. His temperature was 38.5°C and his pulse was 108 with a blood pressure of 90/50 mmHg. His jugular venous pressure (JVP) was elevated and pulsatile. Careful examination of the skin revealed bluish-purple patches.

1 *What do you think is the likely diagnosis?*
2 *Why is the JVP pulsatile?*
3 *Where is the infection likely to be located?*
4 *What is the most likely organism that could be present?*
5 *List the investigations that are necessary.*
6 *What should your initial treatment be?*

Case 2

A 18-year-old female 'gap year' student presents to her general practitioner with sudden onset of fever, myalgia and weakness. She complains of a little diarrhoea and a cough. She has recently returned (1 week ago) from a 3-month-long overland trip in East and Central Africa. She took regular antimalarial prophylaxis throughout her travel overseas. She ate food locally and lived in local hotels and boarding houses. Her temperature is 37.9°C with a pulse of 80 and her blood pressure is normal. Her lung fields are clear and abdomen is soft – there is no evidence of an enlarged liver or spleen. There are the marks of mosquito bites around her ankles but no evidence of skin sepsis.

1 *What significance should be placed on the information that she took her malaria prophylaxis?*
2 *What is the most important infection to exclude?*
3 *How would you do this?*
4 *What other possible diagnoses should you entertain?*
5 *List the investigations that you would order.*
6 *Are there other important steps in the management?*

Case 3

A 32-year-old male presents with fever, headache and myalgia. He works as a green keeper at the local golf course in a semi-rural district. Questioning reveals that over the last few weeks he has been working to refashion the water hazards at the course, spending much of the day in the water. Clinical examination reveals that he has mild neck stiffness. There is a conjunctivitis present and he has a mild jaundice. His abdomen is soft; his liver edge can be palpated and is tender.

The preliminary laboratory findings of his full blood count, urea and electrolytes and liver function tests performed in the accident and emergency department are as follows:

WBC 14.2, predominantly neutrophils
Urea 9.9 mmol/L
Creatinine 138 mmol/L
AST 150 IU/L

1 *What do the initial blood tests show?*
2 *What specific test would you perform?*
3 *What is the most likely infectious diagnosis?*
4 *Are there any additional ways of making the diagnosis?*
5 *What is the significance of his occupation?*
6 *What is the first choice of antibiotic?*
7 *What is the significance of the poor renal function?*

Case 4

A 55-year-old woman who was previously well presents with shortness of breath, fever, cough and a productive sputum. She is a regular smoker and has not travelled abroad recently. Clinical examination reveals that her pulse is 104, her blood pressure 100/60 mmHg and her respiratory rate 32/min. A chest X-ray shows extensive consolidation at her lower left base and a small effusion.

1 *What is your differential diagnosis?*
2 *What is the influence of smoking in this story?*
3 *What further investigations would be helpful?*
4 *A diagnosis of community-acquired pneumonia has been made, but what is the most likely pathogen?*

The patient is treated with ceftriaxone and clarithromycin and there is an initial response to this prescription: the temperature, pulse and respiratory rate fall and she begins to feel better. After 6 days the patient complains that she is feeling worse, and that she is more breathless. The nurses report that her temperature has come back.

5 *What complication may have developed?*
6 *How would you make the diagnosis?*
7 *What additional procedure may be required?*
8 *What microbiological examination should be performed now?*

Case 5

A mother brought her 2-year-old child to surgery complaining that the child had been irritable and had a temperature for the last 2 days. Clinical examination confirmed the temperature and the child was noted to have a runny nose and eyes. Additionally, she had a cough. There was no evidence of rash, and examination of the ears and chest revealed no abnormalities. Advice was given to the mother to use appropriate doses of paracetamol. The mother had one other child who was 5 and neither of her children had received their routine vaccinations because she was concerned about the 'risks of side-effects'.

The on-call service was called that evening because the child had had a convulsion. The attending doctor noted a fine macular rash at the hairline and that the temperature was 39°C, and diagnosed a febrile convulsion.

1 *What vaccines should a child of 2 years have received?*
2 *What is the significance of the febrile convulsions?*
3 *From the clinical picture described, what do you think may be the diagnosis?*

4 *How can the diagnosis be confirmed?*
5 *How could the diagnosis be confirmed by the laboratory?*
6 *Are there any other actions necessary?*

Case 6

Mr S O'D is a 32-year-old male from Ireland who was released early from prison under the terms of the 1997 Good Friday agreement. Some 10 years ago he was the victim of a punishment shooting and since then has suffered from a discharging sinus below his left knee. The clinicians managing the case have always noted that the discharge appeared blood-stained. A number of different organisms have been isolated in the past and he has been given a succession of different antibiotics with varying success.

1 *What is the underlying diagnosis in this case?*
2 *How would you investigate this radiologically?*
3 *What further simple investigations are necessary?*
4 *What other investigations may become necessary should the initial test prove negative?*

The initial sample from pus grows Gram-negative bacilli, the colonies of which appear bright red on the agar. The laboratory identifies these as *Serratia marscescens*.

5 *What is the significance of the isolate?*
6 *What is the significance of the red colour on the plates?*
7 *How should the patient be managed?*

Case 7

A 25-year-old male is a long-standing patient of the renal unit who are managing his end-stage renal failure. He has presented with evidence of recurrent line sepsis that has been treated with antibiotics. A succession of different bacteria have been cultured in the past 6 weeks. On this admission he presents with clouding of vision which is diagnosed clinically as endophthalmitis.

1 *What is the underlying problem for this patient?*
2 *What complications is he prone to?*
3 *What investigations would you recommend initially?*
4 *What more invasive investigation is probably necessary?*
5 *What action will you take at this time?*

A vitreous sample is taken and the culture reveals *Staphylococcus aureus*.

6 *What is the significance of this isolate?*
7 *How will you manage the case now?*

Case 8

A 44-year-old male who presented with an episode of collapse described a 2-week history of fatigue, agitation, confusion and left arm weakness, but no headaches, visual problems, vomiting, seizures or dental pain. On examination he is found to be febrile. He has a reduced level of consciousness.

1 *What is the differential diagnosis?*
2 *What does the absence of papilloedema suggest?*
3 *What investigation is required now?*

A CT scan showed left frontal and right parietal lobe abscesses.

4 *What is the most appropriate treatment?*
5 *What additional treatment should be considered?*

A sample of pus from a brain abscess was sent to the laboratory for microscopy and culture, where a few white blood cells and numerous Gram-positive cocci were seen. Alpha-haemolytic streptococci grew on routine culture media. The isolate was identified by biochemical testing to be *Streptococcus parasanguis*. A 16S rRNA gene PCR followed by sequencing was performed and the isolate was identified as *Streptococcus intermedius*.

6 *How would you interpret these results?*
7 *How would these results affect the treatment?*

Case 9

A 73-year-old women in a care of the elderly ward develops diarrhoea. There is no fever. She has been an inpatient for 3 weeks following a urinary tract infection that has caused a loss of control of her diabetes mellitus. She was treated with amoxicillin for her urinary infection.

1 *What investigation should you send for now?*

The next day the nursing staff report that there are now three other patients on the ward with diarrhoea.

2 *What additional possibilities for the diagnosis are there?*
3 *How should this be investigated?*

The result of the first patient's stool sample comes back and shows that the patient has *C. difficile* toxin.

4 *How does this fit in with the history that you have?*
5 *What is the significance of this result?*
6 *What further investigations should the laboratory perform?*
7 *What needs to happen to control the outbreak?*

Self-assessment case studies: answers

Case 1

1 This is a typical story of a patient with infective endocarditis. The severity and speed of presentation is acute so care must be taken with management.

2 This is because the tricuspid valve is diseased and is incompetent.

3 See answer **2**. You should confirm this clinical diagnosis with echocardiogram.

4 His clinical history indicates that he injects himself and is therefore at risk of infection of his heart valves if his sterile technique fails. With the acute presentation and the history of intravenous substance misuse one naturally thinks of organisms found on the skin and injected. *Staphylococcus aureus* would cause this sort of picture and would be my number one diagnosis.

5 At least two sets of blood cultures, urea and electrolytes (to monitor renal function), full blood picture and C-reactive protein (this may help monitor progression). Echocardiogram including views of the aortic root where abscesses may form. Other investigations may be carried out to detect other infectious risks of intravenous drug use, e.g. hepatitis B, hepatitis C and HIV.

6 Flucloxacillin and gentamicin.

See Chapter 46 for further details.

Case 2

1 It is possible to develop malaria after having taken propylaxis, although it does reduce the risk of infection substantially. Additional history is required to find out what prophylaxis was taken and whether it was appropriate for the region she travelled through. It would also be necessary to find out whether she stopped her tablets immediately after she left the malarious area.

2 Malaria is the most important as *Plasmodium falciparum* infection can be fatal.

3 Thick and thin blood films should be examined by a competent microscopist. There are antigen detection dipstick tests that can be very helpful in making a diagnosis.

4 There is very little to go on in this history and the examination is not very helpful. This is commonly the situation. As she has returned so recently one should exclude typhoid and brucellosis. Sepsis following a urinary infection is possible. There are a wide range of viral infections to consider, including acute hepatitis A. This could also be the 'Katayama fever' stage of acute schistosomiasis.

5 In addition to films for malaria and trypanosomiasis, blood cultures, urinary culture, and faecal cultures should be taken. Serum should be taken and saved for later. A full blood count and white cell count to determine the presence of neutrophils or eosinophils. Viral serology including hepatitis A. Consider HIV testing. Take a faeces sample for detection of parasitic ova and cysts. Serology for parasitic infections including schistosomiasis and filaria.

6 The holiday is now over and the patient may be feeling very anxious. It is also important to call in specialist advice, so this patient should be referred to an infectious diseases or tropical diseases specialist.

See Chapter 40 for further details.

Case 3

1 There is evidence of a neutrophilia, renal dysfunction and liver dysfunction.

2 You will need to investigate infectious causes of hepatitis. Serology for the viral hepatitides should be sent, together with tests for leptospirosis. Surgical conditions such as acute cholecystitis should be investigated by ultrasound examination.

3 This is an interesting case of a man with a relatively rare infection. The findings demonstrate that he has a hepatitis, neck stiffness and conjunctivitis — the classic triad of leptospirosis. The final clue to the diagnosis is the reported exposure to the water hazard, emphasizing the importance of taking a thorough occupational history. The neutrophilia and evidence of renal disorder are also typical of leptospirosis.

4 Leptospirosis can be diagnosed by culture, but this is a difficult technique and few laboratories are able to achieve it.

5 In the past, occupational exposure in sewage workers was a major risk factor for leptospirosis. This is now rare as the workers take appropriate preventive measures. Most leptospirosis is caught by exposure to open water sources during recreations such as wind surfing or fishing.

6 The treatment of choice is benzylpenicillin with doxycycline as an alternative. Treatment should be given as soon as a diagnosis is suspected as delayed treatment is less effective.

7 The prognosis of leptospirosis is defined by the severity of renal damage. Care must be taken in this patient as he has evidence of liver and renal damage.

See Chapter 27 for further details.

Case 4

1 This patient is quite ill so although there are a wide range of possible diagnoses we would be most concerned about infection with *Streptococcus pneumoniae* or *Legionella pneumophila*.

2 Smoking makes infections more likely and tends to produce a worse prognosis than in non-smokers.

3 All of the other causes of acute community-acquired pneumonia should be considered. Thus, sputum should be cultured if available and blood cultures taken. A urinary antigen test for *Legionella* and *S. pneumoniae* is the quickest way of making a diagnosis of Legionnaires' disease and supporting the diagnosis of pneumococcal disease. NAAT tests for the other respiratory bacteria should be performed if available. Virological causes should also be investigated, with real-time NAAT tests if available.

4 *S. pneumoniae*.

5 The improvement that has occurred suggests that the treatment is effective. In this case a pneumococcus was isolated from initial blood and sputum samples that was fully sensitive to penicillin. The relapse of symptoms should prompt a search for one of the suppurative complications, which could include a pleural empyema or, more rarely, a cardiac empyema.

6 Careful clinical examination may reveal signs of an empyema.

7 A diagnostic and therapeutic aspirate of the empyema fluid may be helpful.

8 Aspirated pus should be subjected to conventional culture. Additionally the specimen could be examined by specific NAATs or by a 16S NAAT.

See Chapter 48 for further details.

Case 5

1 This differs from country to country but in the UK children should have been immunized against diphtheria, polio, pertussis, *Haemophilus influenzae* b, tetanus, meningitis C, *S. pneumoniae*, measles, mumps and rubella.

2 Parents are often fearful of convulsions. Although they can be a sign of severe underlying disease, more often their only significance is as an accompaniment to fever. The physician should reduce the temperature and investigate the cause of the fever.

3 Coryza with a febrile convulsion and a fine macular rash at the hairline is a typical presentation of measles. It is notable that the child is unvaccinated.

4 The simplest way to confirm the diagnosis would be to identify Koplik's spots in the mouth of the child. This would make the clinical diagnosis.

5 Laboratories would wish to confirm the diagnosis because of the public health implications. This could be achieved by a saliva sample which could be examined for measles-specific IgM.

6 The contacts have to be considered. Measles is a very infectious condition so it may be necessary to contact the local Health Protection Unit to investigate spread in the community.

See Chapter 32 for further details.

Case 6

1 The discharging sinus is likely to originate in the bone, so a working diagnosis would be chronic osteomyelitis.

2 In chronic osteomyelitis changes in the bone are well established, and thus a plain X-ray should reveal signs of chronic disease. A bone scan or labelled white cell scan is likely to show that there is chronic inflammation at the site.

3 Making an aetiological diagnosis is important here as it is essential to choose the correct antibiotic that penetrates into bone. Empirical therapy would be with flucloxacillin and fusidic acid, but if the patient does not have a staphylococcal or streptococcal infection the natural history of the chronic osteomyelitis will not be interrupted. A sample of pus should be taken for culture and susceptibility testing.

4 In view of the importance of making an aetiological diagnosis an operative sample could be considered. This may have the added advantage of removing infected involucrum.

5 Since this isolate was from a swab that could have been contaminated during sampling the organism isolated must be evaluated for clinical significance. *S. marscescens* is an environmental organism and, given the circumstances of the original injury, could be relevant here.

6 This is a microbiological curiosity — as these organisms are pigmented. This information does help to establish the significance of the isolate, as previous clinicians examining this discharge have found it to be 'blood-stained'. This tends to support the idea that the organism has been present before and strengthens the case for its aetiological role.

7 The patient should be commenced on intravenous antibiotics to which the *Serratia* is sensitive. The first choice would be ciprofloxacin as this will penetrate to the site of infection more effectively than β-lactam antibiotics. Long-term antibiotic therapy coupled with drainage and removal of any involucrum may effect a cure.

See Chapter 50 for further details.

Case 7

1 Patients like this man depend on intravenous lines. The options for siting these lines become more difficult as time goes on so that recurrent line sepsis is often treated conservatively.

2 The patient is at risk of bacterial endocarditis seeded from an infected line tip. Another possibility, and the one most likely here, is that he has developed an endophthalmitis. Specialist advice from microbiologists and ophthalmologists will be required.

3 It is important to establish the degree of inflammation so a full blood picture and white cell count is required, together with C-reactive protein. A set of three blood cultures is required.

4 A sample of vitreous humour. It is important to establish the aetiological diagnosis as this patient may have infection with a wide range of bacteria and penetration to the site of the infection is difficult.

5 The clinical story suggests that this line is infected and should be resited.

6 Given the history of line infection this is a likely pathogen in this circumstance.

7 Treatment should be based on the susceptibility pattern of the organisms. Antibiotics should be prescribed intravenously and subconjunctival injections may be required.

See Chapter 9 for further details.

Case 8

1 Reduced consciousness and fever should make one consider bacterial meningitis, cerebral malaria, brain abscess or cerebral haemorrhage.

2 The history of this presentation is short so the absence of papilloedema does not exclude raised intracranial pressure.

3 A CT scan is necessary to determine whether there is any raised intracranial pressure.

4 The patient should be commenced on a regimen that includes a third-generation cephalosporin and metronidazole.

5 For diagnosis and therapy the abscess could be drained.

6 The molecular identification is more likely to be correct. *Streptococcus parasanguis* is usually found as a commensal in the mouth, whereas *Streptococcus intermedius* is capable of causing abscesses throughout the body.

7 The streptococcus will be sensitive to the cephalosporin and treatment should continue as before.

See Chapter 14 for further details.

Case 9

1 Stools should be sent for investigation of the diarrhoea. The investigation should be focused on the identification of *Clostridium difficile*.

2 There is now a possibility that we have an outbreak of infectious diarrhoea. This may include a food-borne outbreak such as salmonellosis, or a viral infection. Among the viral infections norovirus is a common problem in hospitals.

3 Samples should be sent from all of the involved patients for bacterial culture and viral NAAT.

4 This is what was first anticipated. The previous antibiotics have provided the environment that has allowed the proliferation of *C. difficile* in the gut followed by release of toxins.

5 *C. difficile* may cause outbreaks in hospitals so it is important to find out what the other patients have. The clinician needs to contact the laboratory and ask to expedite the results.

6 The stool should be cultured and the strains typed.

7 There is a need to control antibiotic prescription throughout the ward. Enhanced cleaning of the ward area is required and the patients should be isolated. The importance of hand-washing for all staff should be re-emphasized.

See Chapter 18 for further details.

INDEX